9/2000

The Limits of a Limitless Science

and Other Essays

By the same author

Les tendances nouvelles de l'ecclésiologie

The Relevance of Physics

Brain, Mind and Computers
(Lecomte du Nouy Prize, 1970)

The Paradox of Olbers' Paradox

The Milky Way: An Elusive Road for Science

*Science and Creation: From Eternal Cycles
to an Oscillating Universe*

*Planets and Planetarians: A History of Theories
of the Origin of Planetary Systems*

The Road of Science and the Ways to God
(Gifford Lectures: University of Edinburgh, 1975 and 1976)

The Origin of Science and the Science of its Origin
(Fremantle Lectures, Oxford, 1977)

*And on This Rock: The Witness of One Land
and Two Covenants*

Cosmos and Creator

Angels, Apes and Men

Uneasy Genius: The Life and Work of Pierre Duhem

Chesterton: A Seer of Science

The Keys of the Kingdom: A Tool's Witness to Truth

Lord Gifford and His Lectures: A Centenary Retrospect

Chance or Reality and Other Essays

The Physicist as Artist: The Landscapes of Pierre Duhem

The Absolute beneath the Relative and Other Essays

The Savior of Science
(Wethersfield Institute Lectures, 1987)

Miracles and Physics

God and the Cosmologists
(Farmington Institute Lectures, Oxford, 1988)

(continued on p. 248)

The Limits of a Limitless Science

and

Other Essays

by

Stanley L. Jaki

ISI Books
Wilmington, Delaware
2000

Cataloguing-in-Publication Data

Jaki, Stanley L.
      The limits of a limitless science : and other essays / by Stanley L. Jaki.—1st ed.—Wilmington, DE : ISI Books, 2000.

    p. cm.

    Includes index.
    ISBN 1-882926-46-3
    1. Science. 2. Science—Philosophy. I. Title

Q171 .J35 1999   99-68034
500—dc21   CIP

ISI Books
P. O. Box 4431
Wilmington DE 19807-0431

Cover design by Glenn Pierce

Printed in the United States of America

Contents

Introduction

For all the variety of their titles, the chapters of this book convey a concern common to all of them. It is the concern about the importance of knowing the limits of the scientific method, a concern which no less a physicist than James Clerk Maxwell specified as a most difficult task for a scientist to cope with. Maxwell clearly had much more in view than a purely theoretical task, although, if anyone, he knew the enormously vast bearing of what is shown to be theoretically right. Truth, which has its intrinsic value regardless of application or action, controls in the end all pursuit, scientific or other.

Therefore, since no tool used by man matches even remotely the effectiveness and range of the tool called science, one may rightly say that there is nothing so important as to ascertain the limits to which science can rightfully be put to use.

In Chapter 1 a rule of thumb is offered whereby even a non-scientist can readily perceive something of the limits of science, which in a sense is limitless in its applicability. Since quantities are everywhere where there is matter, science claims its rights whenever one confronts matter whether on the very large or on the very small scale, whether very far from man or in man's very interior.

But science ceases to be competent whenever a proposition is such as to have no quantitative bearing. The alternatives—to be or not to be, to be free or not to be free, to act for a purpose or to act for no purpose at all, to have inalienable rights or not to have them—cannot be evaluated in inches or ounces, in volts or in amperes, in frequencies or in wavelenghts. A brief reflection on this point may reveal even to the non-scientist that the limits of science are vast as well as very specific.

There are, of course, limits of science that are only apparent. The search for extraterrestrials and for other planetary systems may not forever remain without success. But, as is argued in Chapter 2, it is not at all scientific to ignore grave and so far unsolved scientific problems, such as the origin of our planetary system and of our earth-moon system, when one appraises the chances of that success.

Basic limits of the scientific method are, however, once more in view in Chapter 3 that deals with the question of artificial intelligence. It is argued there, and with no reservations whatsoever, that the construction of a computer that thinks and does so in the sense in which man always does, that is, consciously and with a commitment or love, is absolutely impossible.

A historical type of the limitations of science is put in view in Chapter 4, which is about the biblical basis of the rise of science. It is shown there that some non-scientific, in fact a metaphysical and theological input was necessary for the only viable birth of science, which took place in the Western World, and there alone.

Chapter 5 deals with a limitation of science that refers to the inspiration or sense of purpose which it cannot supply, although science, in particular the science of astronomy, has been often presented as the source of that kind of inspiration. This abuse of astronomy has all too often issued in a rank counter-inspiration, best to be called plain despondency or despair.

A most fundamental form of the limitation of the scientific method is presented in Chapter 6. This form relates to the spatial representation one can give to the range of the meaning of various words. It is argued there that only words corresponding to numbers (integers) can be given a geometrical representation. Since most words convey non-quantitative meaning, human language is essentially non-scientific, that is, quantitatively imprecise, and therefore its precision, the basis of all thinking it may have, must point far beyond the limits of science.

It is the consideration of this non-quantitative feature of much of human language that carries one most effectively to realms well beyond science as argued in Chapter 7. Within those realms is located even that largest conceivable material entity, the universe itself, the topic of Chapter 8. But, as shown in Chapter 9, even the investigation of small bits of matter, such as meteors, can reveal telling limitations of the scientific method.

Chapters 10 and 11 deal with those limitations with respect to modern scientific cosmology, insofar as it takes the universe for its real target of investigation. This target remains just as unattainable for science as is a definitive theory of fundamental particles. To purse such a theory is to pursue a dream.

Science nowhere shows so thoroughly its limitations, and these in turn are nowhere more systematically ignored, than in dealing with the relation of science and religion, the topic of Chapters 12 and 13. The last two chapters are in some sense autobiographical. They give a few glimpses of the author's now forty-year-long engagement with the problem of the limitations of science so that human life may become both enriched by science as well as unfettered by it.

These essays, written over a period of seven years, contain some repetitions. Their presence may further bring out pivotal points and arguments. These cannot be repeated often enough if one is to find one's bearing in a world which, though increasingly that of science, cries out for a vivid grasp for its built-in limitations so that fully human perspectives may remain in sharp focus.

August 1999   S. L. J.

1

The Limits of a Limitless Science

The title of this essay is paradoxical and intentionally so. It aims, as do all paradoxes, at alerting the mind to something that should be obvious, but is not necessarily so. For little attention is paid to the obvious fact that whatever one's idea about science, it does not reflect a consensus about it. In fact there are almost as many ideas about science as there are philosophers of science. But there is at least one aspect or feature of science which must be included in all ideas about it. Indeed, as will be seen, that feature guarantees a limitless application of science or of the scientific method, and, at the same time, constitutes its most tangible limit. The feature relates to the fact that for a proposition or reasoning to qualify as science, it must be subject to being tested in the laboratory, or, in general, by scientific instruments.

This does not mean that a reasoning which cannot be tested in the laboratory is unreasonable. It is not possible to dispute, without

First published in *Asbury Theological Journal,* 54 (Spring 1999), pp. 23-29; reprinted with permission.

waxing philosophical, that philosophy is not a reasoned discourse. Theology, as cultivated within Christian circles, was long ago proposed as a science by no less intellects than Augustine and Thomas. But they and an army of lesser though still eminent intellects called philosophy and theology a science only because they saw in it a discourse consistent with its basic premises and presuppositions. It is in that sense that they took philosophy and theology for *scientia*. One wonders whether they would insist on that name nowadays when the word *science* has a very special meaning attached to it. That special meaning is tied to the word laboratory and the kind of work done there.

For although the word laboratory may, etymologically, mean any place where one works, it actually denotes a place where one works for one single purpose: to make observations or measurements which are accurate, so that accurate predictions may be made on their basis. Science, in that sense, is synonymous with measurements, which are accurate because they can be expressed in numbers. Those numbers relate to tangible or material things, or rather to their spatial extensions or correlations with one another in a given moment or as time goes on. All the instruments that cram laboratories serve the accurate gathering of those numbers, or quantitative data.

This is most evident with respect to physics, the most exact form of what in modern times is called science. Any branch of physics is an example of this. There is a science of dynamics because something about what is perceived as attraction among bodies can be measured. There is a science of acoustics because the intensity and speed of sound are measurable. There is a science of optics because a great many things can be measured about light rays, such as the diminution of their intensity with distance, and their ways of propagation. There is a science of electricity, because one can measure the magnitude of electric charges and the forces of attraction and repulsion among them. There is a science of electromagnetism because it is possible to measure the interaction of electrical and magnetic forces that generate electromagnetic waves. These in turn have velocity, amplitudes, frequencies, and a number of other parameters that can be measured. There is atomic and nuclear physics because atoms and their nuclei have measurable properties. Astronomy is a science because the size of stars, their distances from one another and the processes within them can be so many objects of measurement.

Outside physics, all branches of science have tried to emulate physics by restricting their work as much as possible to measuring. In that respect chemistry has achieved a status of exactness practically equivalent to that of physics. Great advances have been made toward exactness in biology ever since Harvey made measurements that revealed the circulation of the blood through the body. Every page of modern molecular biology and biophysics attests to the overriding importance of measurements.

In very recent times psychologists, historians, sociologists, and economists claimed for their studies a scientific status. One speaks today even of political science, or the science of politics. The reason is that in all those fields a great deal of systematic data gathering can take place on the basis of which much can be said about future developments either in individuals or in society. But whenever perfect accuracy is claimed for such predictions, the free character of individuals and of society, together with the freedom of scientific research, is called, at least implicitly, into doubt. At any rate, the gathering of data and measurements in those fields never achieved a predictive accuracy which is on hand in physics, astronomy, and chemistry. In all these fields, science is practically synonymous with the very special work done in laboratories. The work is to take measurements.

Physicists have expressed in various forms their awareness of the importance which measurements have in their efforts. In doing so they have also voiced crucial points about what is involved in making measurements. One such point is the scientific status that can be given to a branch of investigation insofar as it includes measurements. In fact, at times they spoke as if knowledge not based on measurements was mere "hot air," to recall Lord Rutherford's aside to Samuel Alexander, a well known metaphysician.[1] At other times they merely hinted at the lower status of non-scientific knowledge. Such was the case when Lord Kelvin, in discussing in 1883 the electrical units of measurement, recalled a favorite comment of his on what constitutes science: "I often say that when you can measure what you are speaking about, and express it in numbers, you know something about it; but when you cannot measure it, when you cannot express it in numbers, your knowledge is of a meagre and unscientific kind: it may be the beginning of knowledge, but you have scarcely, in your thoughts, advanced to the stage of science, whatever the matter may be."[2]

There is no point in speculating to what extent Lord Kelvin wanted to slight thereby other sciences and non-scientific branches of knowledge. Even in physics it was true that a branch of it, or any theory indeed, increased in scientific value in the measure in which it involved measurements, that is, numerical values. The most concise expression of that view is due to Robert Mayer, one of those who put the mechanical theory of heat on a firm footing. His dictum, "In physics numbers are everything, in physiology they are a little, in metaphysics, they are nothing,"[3] also states a parameter along which various branches of learning can be seen as qualifying in various degrees as science.

Mayer, a physician by training and profession, would, of course, find himself contradicted today by the role of numbers, that is, exact measurements, in physiology. His remark on metaphysics could be justified only insofar as numbers were taken to be equivalent to making measurements. Obviously, ideas cannot be measured by calipers or dissected by microtome. However, insofar as numbers represented the metaphysical category of quantities, Mayer should have known that the best metaphysicians have always viewed the category of quantities as being above all other categories. More of this later. Still, Mayer was correct in saying that in physics numbers were everything. In reference to physics he was fully justified in declaring: "One single number has more real and permanent value than an expensive library full of hypotheses."[4] He had in mind his many years of labor to establish as precisely as possible, and in different circumstances, the mechanical equivalent of heat.

To that kind of labor, done in laboratories, physicists have often assigned the status of judiciary tribunals for science. Einstein had in mind certain numerical values implied in his general theory of relativity when he said in 1919 that "if a single one of the conclusions drawn from it proves wrong, it must be given up; to modify it without destroying the whole structure, seems to be impossible."[5] The experimental evidence that the separation of the components of the fine structure "doublet" in H_α is only 96 percent of that predicted by Dirac's relativistic quantum theory, forced the working out of quantum electrodynamics. Many other examples are provided by classical and modern physics about the supreme role played by measurements.

No wonder that physicists could speak of their being devoted to measuring things and processes as if it had been an obsession with them. A. Kundt, a physicist famous for his measurements of the velocity of sound in gases and solids, once stated that "in the end one might just as well measure the velocity of rainwater in the gutter." What he meant, of course, was not a commitment to measuring for the sake of measuring, a sort of a scientific *l'art pour l'art*. He meant the establishment of a numerical figure that forms part of a meaningful scientific theory. It was in that sense that F. Kohlrausch, a physicist also famous for his painstaking as well as decisive measurements, found Kundt's suggestion congenial to his own way of thinking: "I would be delighted to do so"[6]—was his comment.

Science is competent wherever and whenever the object of investigation offers a quantitatively determinable aspect. The range of science is not limited either by the dimensions of quarks or by the distance to the farthest galaxies. Science touches on all matter—whether solid, liquid, gas, plasma, or a mere flow of energy waves—insofar as matter is extended and therefore measurable. Consequently, science is applicable wherever there is matter in any form whatsoever, because all such matter has quantitative parameters. In that sense science is limitless and its statements are unlimitedly, that is, universally valid throughout the universe of matter.

It is another point that the total amount of matter, which is measurable, has to be finite. An infinite amount of measurable matter would be the embodiment of an actually realized infinite quantity which is, of course, a contradiction in terms. Again, it is another matter that the quantitative ascertaining by science of the quantitative structure of matter can never have an end to it. A physical theory can never be final for two reasons. For one, physicists can never be sure that they will never stumble on some previously unsuspected features of matter. Also, even if they were to succeed in formulating a "final" theory, they can never be sure theoretically whether it is really final. For as long as Gödel's incompleteness theorems are valid, the mathematical structure of that theory cannot contain within itself its own proof of consistency.[7]

Science, insofar as it deals with quantities, is not limited by non-quantitative considerations. No non-quantitative set of considerations, be they metaphysical, theological, or aesthetic, can set a limit to the competency of science. But this limitless character, which science enjoys with respect to the quantitative aspects of reality, is also the source of its drastic limitedness. This is clear even within that most quantitative of all sciences which is physics. Thus to take only one branch of physics, electromagnetism, its scientific status rests on the fact that electromagnetic waves have measurable properties. Perplexities envelop that status no sooner than one yields to logic by saying that if there are waves, electromagnetic or other, there has to be something which performs that wave-like motion. For if that something is called electricity, an answer is given that appears satisfactory until one asks: what is electricity?

An answer to that question evades the physicist. This was admitted even by Lord Kelvin, who stuck to the end to his conviction that ultimately every physical process and property had to be mechanical. Yet this was not what he said to a young foreman in a big Glasgow electrical equipment factory, who happened to guide through the plant the great physicist, without recognizing who he was. After Lord Kelvin listened with great patience to elementary instructions about condensers, insulators, magnets, and whatnot, he decided to give a gentle lesson to the all-knowing young man, by asking him: "What is electricity?" On finding, not surprisingly, that the young man was at a loss for words, Lord Kelvin gently assured him: "This is the only thing about electricity which you and I do not know."[8]

Physicists are, of course, apt to answer that electricity is a field, like gravitation. But then another question can be raised: What is a field? If we answer that a field is an oscillation, the answer begs the question: what is it that oscillates? A particular difficulty arises if one goes on to quantum electrodynamics with its fields of zero-point oscillations in the vacuum. Can such a vacuum be really empty, as it should if it is truly a vacuum?

The quantum-mechanical vacuum and its no less baffling conceptual affiliates are but the latest in the long list of such questions. Already in Newton's time it was realized that physics cannot answer the question: what is gravitation? Physics could only establish the fact that what is called gravitation has a quantitative property, known as the inverse square law. Almost two hundred years later, when Maxwell worked out his electromagnetic equations, much effort was spent, but in vain, on finding out what it was that acted according to those equations. The most popular theory, indeed conviction, was that the something in question was the ether, but physicists could not measure anything about it. They could not even measure a presumed effect of it on the speed of light beams sent through it.

This famous null-result of Michelson's experiments forced physicists to declare the ether to be non-existent. Indeed this conclusion of theirs was valid insofar as the ether was thought of as something material and therefore an entity with measurable properties. Even today, after all the successes of quantum electrodynamics, physicists cannot improve on Heinrich Hertz, the first to demonstrate effects predicted by Maxwell's theory, effects naturally interpreted as produced by electromagnetic waves. In an almost despairing tone Hertz wrote: "Maxwell's theory is Maxwell's system of equations."[9] Such is a concise expression of the radical limits put on science to say anything about the so-called electromagnetic medium, be it called field or vacuum or whatnot.

Maxwell's equations, like all other equations of physics, are a set of quantitative correlations. Nothing more, nothing less. And the same can be said of the tensor equations of the general theory of relativity and of the matrices and wave equations of quantum mechanics. In other words, if Hertz's remark is right, one can say that Einstein's general theory of relativity is Einstein's system of equations, or a set of generalized quantitative correlations. Indeed, one may say that all theories of physics are generalized sets of quantitative correlations.

This conclusion sets very sharp limits to the applicability of science: wherever reality offers aspects with no quantitative properties to be measured, science is not applicable. In addition, as will be seen, the scientific specification of those quantitative properties cannot be taken for an initial instalment on specifying non-quantitative properties of the same reality. Quantities forever remain quantities, conceptually that is. This is, however, not something to trouble scientists insofar as they assume, on the basis of commonsense wisdom, that there are things and processes to measure and they are satisfied with measuring them. But the same restriction of the applicability of science keeps troubling some scientists. These, being overawed by the success of the quantitative method, think that science should be applicable in every field of human experience and reflection.

This conviction of theirs can manifest itself in an almost incidental, yet very startling manner. A good example of this can be found in a chapter which deals with "The Flow of Dry Water," that is, non-viscous flow, in the *Feynman Lectures,* a highly regarded textbook on physics. There Feynman makes two final remarks. One is that from the relatively simple principles governing non-viscous flow an "infinite variety and novelty of physical phenomena . . . can be generated; . . . we just haven't found the way to get them out."[10] This may be taken merely for an ambitious program for physics, provided one does not take the words, "physical phenomena . . . can be generated" for an endorsement of some ill-digested form of Platonism where numbers produce physical reality.

But a worse perspective, that has nothing to do with physics, transpires from Feynman's next remark in which he bemoans the fact that "today we cannot see whether Schrödinger's equation contains frogs, musical composers, or morality—or whether it does not."[11] Feynman does not say categorically that it does. Yet by taking as plausible the possibility that Schrödinger's equation may contain all that, Feynman claims that science is limitless in a sense very different from the one already stated, namely, that science is applicable wherever there are quantitative properties to measure. This unlimited-ness of science is extended by Feynman into a sweeping suggestion with no restriction whatsoever: Not only matter but everything else, morality included, can be measured, and indeed is contained in some future form of physics.

What Feynman put forward in an almost incidental way, other prominent physicists present in a systematic manner. An example is Roger Penrose's book, *The Emperor's New Mind*. There he argues that some new, and so far unknown form of general relativity, which is quantized and therefore statistical, will contain the full explanation of all human thought. There is much more to Penrose's idea than the far from demonstrated claim that thoughts can be measured. What Penrose really claimed was the old Platonic idea that ideas of quantities necessarily turn into real matter with quantitative proper-ties. Therefore, since mathematical physics is the best way of dealing with the quantitative properties of matter, mathematical physics is declared to be all that we need in order to cope with existence, material as well as intellectual and moral.[12]

A variation on this claim is found in Stephen Hawking's book, *The First Three Minutes*. According to its grand conclusion, a theoretical cosmology, which is so perfect as to be free of boundary conditions, automatically assures the existence of the universe.[13] Another prominent physicist, A. H. Guth of MIT, relies even more crudely on this rather naive cavorting, in the name of physics, with Platonism. According to him quantum cosmology gives the scientist the power to create universes "literally" and "absolutely out of nothing."[14] If, however, physics turns the physicist into a Creator, there remains absolutely no limit to science.

Now, if such scientifically coated claims are true, one might as well follow the advice which David Hume gave at the end of his *Enquiry concerning Human Understanding* and burn all books except those that contain quantities and matters of fact.[15] Obviously, Hume meant only those facts that were material and therefore could be measured or evaluated in terms of quantities. At any rate, ever since Hume the book burning recommended by him has been busily done, at least in a metaphorical sense. It is, however, well known that during the French Revolution and kindred ideologico-political revolutions, pyres were made of books that lacked quantities and matters of fact as understood by Hume. Metaphorically, that book burning can be done (and this is the way it is done in the name of science), by declaring that anything that cannot be measured is purely subjective, almost illusory. Einstein himself claimed that since our experiencing the "now," which is the very center of human consciousness, cannot be measured, it is a purely subjective matter.[16] He said the same thing about free will as well.[17]

But then should a scientist accept a prize, say the Nobel Prize, for his work if he was not really free when he worked for his discoveries? Are we to reward sheer automata with huge and prestigious awards? But how would this be a non-automatic process? In order to have a proper answer, one should first recall the penetrating observation of a scientist, Henry Poincaré, who, about hundred years ago, called attention to an elementary fact: "Even a determinist argues non-deterministically."[18] Clearly, both within and outside science, the claim that science has no limits would, if rigorously applied, lead to absurd consequences.

Nothing would remain of the criminal justice system, if a criminal could claim, with a reference to Einstein, that his consciousness of having committed this or that crime was purely subjective. Should criminal justice courts admit purely subjective evidence and impose, on that basis, huge prison terms?

If free will is purely subjective, what becomes of arguments, either scientific or philosophical, against free will? Those arguments then cannot be submitted as having objective validity. But an argument which is not the result of free deliberations, is not an argument. If, however, those arguments against the freedom of the will are not purely and blindly mechanistic processes, they are so many proofs of the existence of free will.

Such consequences are at times blissfully ignored by prominent scientists. Worse, they fail to notice that even in their own scientific field mere measurements and quantities do not suffice. A recent illustration of this comes from the Einstein memorial lecture which Professor Watson, the co-discoverer of the double helix structure of DNA molecules, gave at Princeton University on February 16, 1995. There he amplified his statement that all human life can be described in terms of molecules, with the words: "There is no need to invent anything else."[19]

Should one then say that life as such does not matter, because it cannot be observed or measured, in spite of the enormous successes of biochemistry, biophysics, microbiology, and genetic research? They certainly show the enormous extent of measurable parameters in life processes. But life itself still cannot be measured. Therefore, scientifically speaking, life does not exist. This paradoxical fact was in the mind of Claude Bernard, the great French physiologist, when he made, around 1860, one of his famous statements. On being asked whether, in reference to life, he was a vitalist or a mechanist, he replied that he had never seen life.[20] This was his way of calling attention to the fact that by becoming either a vitalist or a mechanist, one moved beyond science. There remains, indeed, much more to the question, What is Life? than can be dreamt of by biochemists or biophysicists who take the mechanistic outlook on life. Equally, biologists who espouse vitalism are dreaming when they imply that they can see experimentally purposiveness, this chief characteristic of life processes. Just as the mechanistic interpretation of life is a philosophy, so is vitalism. Both are bad philosophies, though in opposite sense. In the former the claim is made that just because something (purposiveness) cannot be measured, it does not exist. In vitalism the claim is made that somehow purposiveness can be measured and therefore becomes part of experimental science.

Contrary to the claim that DNA is the secret of life, life remains the secret of DNA. Microbiology has not found a quantitative answer to the apparently purposeful action in all living things, from cells to mammals. Microbiology has not found a quantitative answer to the question of free will. Brain research cannot answer the question, What is that experience, called "now," which is at the very center of consciousness? For even by finding the exact biochemical conditions that are connected with the personally felt consciousness of the "now," the question what is that "now" remains to be answered. While brain research may establish the biochemical processes whenever a given word is thought of, it cannot account for what it is for a word to have a meaning.[21]

Faced with that inability, the scientist can take two attitudes. One rests upon the mistaken conviction that the scientific method is everything and whatever cannot be expressed in quantitative terms, is purely subjective, that is, illusory. Such was, for instance, the attitude of Einstein, who said that consciousness and free will are no objective realities, because they cannot be handled by physics. He might as well have called them sheer illusions. Clearly, it is better to take another attitude and acknowledge that there are some basic limits to a limitless science. Those limits appear as soon as a question arises that cannot be put in a quantitative form and therefore cannot be given a quantitative answer to be tested in a laboratory.

Such are indeed all the major questions of human existence. To answer the question, "To be or not to be?" we cannot turn to a science textbook. Strictly speaking, for Hamlet the question meant a choice between two courses of action: one was to continue to live by ignoring an immoral situation. The other was to take revenge and run thereby the risk that one's physical being would come to an end. Already that moral choice demands far more than some quantitative testing in a laboratory.

But Hamlet's question, "to be or not to be," has a meaning even deeper than whether an act is moral or immoral. That deeper meaning is not merely whether there is a life after death. The deepest perspective opened up by that question is reflection on existence in general. In raising the question, "to be or not to be," one conveys one's ability to ponder existence itself. In fact every bit of knowledge begins with the registering of something that exists. To know is to register existence. But this is precisely what science cannot do, simply because existence as such cannot be measured. Yet, worse than impotency is on hand in thinking about existence when it is done in terms of a philosophy that apes scientific parlance. A philosopher, no matter how eminent, makes a mockery of his field when he keeps asserting that "To be . . . is to be the value of a variable."[22] Compared with this, it may strike one as an innocent joke to say, as did a graduate student of physics, that Hamlet was an internal combustion engine, with a very low efficiency, because he could have prevented the death of six people by simply killing one, his mother.[23]

When science establishes, for instance, the quantitative knowledge that the earth's diameter is so many kilometers, it presupposes first of all the fact that something, the earth, does exist. And this holds true of any quantitative result of science, such as atomic radii, distances to other galaxies, the characteristic wavelength of the cosmic background radiation etc. Atoms and galaxies are useful objects for science because they exist in such a way as to have quantitative properties. Moreover, science can, by refining more and more its measurements of those properties, establish such sets of them that are conveniently called new subatomic entities.

But it would be mistaken to assume, although this is customarily done, that science, or rather its quantitative method, finds new entities in the ontological sense. Science merely uncovers new aspects in the vast gamut of material existence. Were it otherwise, one would endorse the Platonic fallacy that it is the quantitative properties that give existence to material entities. Moreover, were such the case, nothing would exist that cannot be given a quantitative formulation. In that case such words as conscience, free will, purpose, moral responsibility, to say nothing of the soul, would be so many empty words, standing for anthropomorphic illusions. But, there would not be, in that case, even scientific investigations, because there would be no scientists who would investigate things freely and be conscious of the fact that they are investigating.

In other words, there is a most fundamental limit to a limitless science. Science has no limits whenever it finds—and in whatever form—matter or material properties. There is no limit, for instance, to measuring the physiological processes that take place in the brain. when one thinks as much as a single word. It is possible that one day brain research will be so advanced and exact as to give a complete quantitative account of all the energy levels of all the molecules in the brain when one makes the conscious reflection on the "now." But even then there remains the radically non-quantitative character of that experience, a character clearly recognized by Einstein. He merely failed to recognize the limits of science when he stated that whatever cannot be measured and therefore be expressed in quantitative terms, cannot be objectively real.

Einstein failed in this respect, as did and do so many other prominent scientists. He failed a test to which his great idol, James

Clerk Maxwell, gave the best formulation. Towards the end of his most distinguished career, Maxwell put in print the following words: "One of the severest tests of the scientific mind is to know the limits of the legitimate application of the scientific method."[24] That method can be legitimately applied wherever one can find experimentally verifiable quantities.

The expression, "experimentally verifiable quantities," is crucial. Theology too deals with quantities. Theology states that there is one divine nature in three divine persons; that in Christ two natures exist in one person; that there are seven sacraments and four canonical gospels. But none of these numbers can be tested in laboratories. Nor can there be a laboratory test of the philosophical and theological tenet that the world was created out of nothing. There can be no such test because the nothing, insofar as it is radically no thing at all, cannot have observable and therefore measurable properties. For the same reason science cannot establish the first moment of cosmic existence. For to prove that moment to be truly first, one has first to show that nothing preceded it. Such is, however, a purely negative proviso that cannot be given a quantitative, that is, scientific verification. Nor is the universe as such an object for science. Scientists cannot go outside the universe in order to observe the whole of it and thereby give to their knowledge of the universe that supreme scientific seal, which is observation with measurement.

Moreover, the notion of the universe is the vastest of all generalizations, although in more than a purely quantitative sense. The notion of the universe is the supreme form of universals. Quantitative or experimental considerations, that always relate to the particular, remain wholly insufficient to justify the validity of the universals, including above all the universe. Indeed, no branch of modern science, with one exception (evolutionary biology), is so fundamentally dependent on philosophy as is scientific cosmology, and in no other field of science is philosophy more ignored, and indeed scorned.

In scientific cosmology, insofar as it deals with various components of the universe, such as galaxies and globular clusters, philosophy can be safely ignored. The scientist merely has to assume, on a commonsense ground, that those objects do exist because they are observable. Only when it comes to the universe as such, do scientific cosmologists claim to know something whose existence only a rigorously articulated philosophy, respectful of the universals, can demonstrate.[25] But in evolutionary biology one comes across indispensable philosophical terms at almost every step. Concerning the species, it is something that cannot be observed. Yet it has to exist if it is right to talk about the origin and transformation of species. One can get around this problem, which involves the philosophical problem of knowing universals, by defining a species as the totality of all individuals that can interbreed. But when we go to the genus and to even higher units, up to families, orders, phyla and kingdoms, that definition does not do. Again, only the great generalizing powers of the mind can enable the evolutionary biologist to see a continuous connection along the paleontological record, although, as recent findings show, it is more riddled with huge holes and discontinuities than ever suspected.

Yet most evolutionary biologists have only contempt for philosophy, although it alone can justify their great unifying vision, which is much more than science, strictly speaking. What they do is climb the rungs of an essentially philosophical ladder in order to see much farther than would be allowed, strictly speaking, by the data on hand. However, once at the top of the ladder, they haughtily kick it away. In doing so they follow the example set by Darwin. With Darwin they try to discredit philosophy with their science, although philosophy enabled them to raise their eyes to heights where biological evolution can be seen, though only with the eyes of the mind. No wonder that the present-day perplexity of some leading paleontologists evokes the fate of Humpty Dumpty.

But just as scientists cannot ignore philosophy, philosophers and theologians can only at their gravest peril ignore the fact that quantitatively verifiable parameters as such lie outside their competence. Herein lies a basic limit of theology, philosophy, and various branches of the so-called humanities. The truth of any philosophical and theological statement that contains experimentally verifiable quantities, depends on experimental or laboratory verification, with measurement being its very gist. This verification only science can provide. Conversely, genuinely philosophical statements cannot have a scientific verification, which always has to be experimental and therefore quantitative, derived from measurements.

Experimentally verifiable quantities represent a basic demarcation line between the sciences and other forms of reasoned discourse, such as philosophy, theology, and so forth. In essence this point was concisely stated already in the *Categories* of Aristotle. This is not to suggest that there is strict logic in his listing number and speech (syllables) as "discrete" quantities and space and time as "continuous" quantities. But there is a perennial truth in his observation that there is no common boundary between two numbers. Nor is it possible to dispute his statement that "the most distinctive mark of quantity is that equality and inequality are predicated of it." Consequently, within the Aristotelian perspective numbers do not admit contraries such as the ones that occur between celestial and terrestrial motion, as well as among the basic types of motion (upward and downward) below the orbit of the moon. Moreover, "quantity does not admit of variation of degree. One thing cannot be two cubits long in a greater degree than another. Similarly with regard to number: what is 'three' is not more truly 'three' than what is 'five' is five."[26]

Leaving aside Aristotle's dicta on space and time, let alone syllables, it should be clear that equality and inequality are, unlike numbers, not absolute, but relative properties that reflect our judgments of similarities among various things. These similarities cannot be translated into absolutely valid numerical propositions. To quote Aristotle: "That which is not a quantity can by no means, it would seem, be termed equal or unequal to anything else. One particular disposition or one particular quality, such as whiteness, is by no means compared with another in terms of equality and inequality but rather in terms of similarity."[27] But it is precisely this kind of similarity that does not lend itself to strict, invariably valid, numerical evaluation.

Aristotle's own examples are worth recalling. A mountain, though a huge entity, can be called small; and a grain, though puny, large. In both judgments comparisons or similarities are at play. Similarly, it is possible to say that a house has many people in it, whereas a theater only a few. All this is but an aspect of what Aristotle specifies, in taking up the discussion of qualities, as the rule of "more or less." Whereas all qualities can be presented as containing "more or less" of what is distinctive of them, this cannot be said of numbers. Aristotle did not suspect that with the coming of science in the modern sense, that is, a science in which quantities rule supreme, this quality of "more or less," so characteristic of his own physical science, would have to be jettisoned.[28]

While Aristotle correctly specified the most important feature of quantities, he himself did not pay proper attention to it as he set forth his accounts of the celestial and terrestrial world. There his "qualitative" physics led him, time and again, into implicitly quantitative, and at times explicitly quantitative inferences that could not be reconciled with what was plainly observed. Thus, it should have been obvious in Aristotle's time that bodies of greatly differing weights fall to the earth in remarkably equal times. Peripatetic physics could have indeed been held up to well-deserved ridicule long before Galileo's time.[29] It was another matter whether it was right for Galileo to invoke Plato's reification of numbers in order to justify that equality and other geometrical or numerical features discoverable in the physical world.

Unfortunately, Aristotle did not set a pattern as to the conclusions to be drawn from what he so incisively stated about quantities. Hence the pathetic opposition posed by Averroist Aristotelians to the new physics of motion. To be sure, even three centuries later one could still be puzzled by the fact that the law of inertial motion, in which there is at least the continued novelty of spatial displacement, "worked," although it provided no "explanation" of how the novelty came about.[30] Yet once one accepted with Aristotle the special status of numbers or quantities, it should have been possible to state that spatial displacement, as a purely quantitative proposition, implied no ontological factors. Not that these factors were denied by the mathematical formalism of inertial motion; rather, mathematics could have no bearing on them. This is no less true of the Newtonian law of accelerated motion. For that law too is independent of whatever philosophical or ontological definition one gives to the force being constantly at work in order to make real the acceleration. For what Hertz said about Maxwell's theory of electromagnetism, can also be stated of Newton's theory of gravitation: it is Newton's system of equations. Nothing more, nothing less.

But it still goes against the grain to recognize that there is an insurmountable conceptual obstacle to the age-old striving after a unified form of knowledge. That obstacle lies in the way of any form of reductionism, crude or refined, vicious or well-meaning. This is not to suggest that there will soon be an end to efforts that want to reduce man, and all his thinking and volitions, to a machine. Scientific reductionism is as strong, if not stronger, than ever. But that obstacle also vitiates all efforts to "integrate" science with philosophy and theology. For if pathetic is the claim that qualities—in their broadest sense, that is insofar as they include value- and existence-judgments—can be reduced to quantities, so is the effort to wring some theological truth out of any result of classical or modern physics. The reason lies in what Aristotle had already stated concisely about the unique status, among all categories, of the category known as quantity.

There is indeed an ineffaceable line of demarcation between the conceptual domain of quantities and all other conceptual domains taken together. The conceptual domain of quantities is a most special domain that stands apart from the rest because all terms belonging to it have a peculiarly common characteristic. They all are strictly univocal, to use a term that, though increasingly unfashionable, never becomes antiquated. Quantities are like so many building blocks with well-defined contours. The co-ordination of those blocks can, of course, be exceedingly complex and complicated. This is brought out by a mere look at any advanced textbook of mathematical physics. But whatever that complexity and complicatedness, the art of handling those blocks—quantities—follows invariable rules. The art is always the same, because the art is a skill with the basic operations of arithmetic.

One can, of course, philosophize about quantities, but the operations performed with quantities have their own independence, precisely because quantities have a specific conceptual character that makes them distinct from all other concepts. All conceptualizations of numbers can be compared to strictly defined building-blocks that remain forever identical to themselves. This is why their addition, subtraction, multiplication, and division are a straightforward matter, in principle at least. All other concepts are amoebas by comparison, or even less definite. For amoebas, although they constantly change, have a strict boundary membrane. Non-quantitative concepts have neither permanent shapes, nor distinct envelopes. This is why their definitions given in any dictionary are forever subject to slight and at times drastic rewriting.[31] This is what Whitehead wanted to convey in warning against the dream of a "Perfect Dictionary."[32]

What is actually being done in giving the dictionary definition of a word is to define it in terms of some other words. In the case of quantitative concepts, say a number, such as twenty, one juxtaposes so many unit areas. Thus the concept of twenty is the sum of twenty units. To represent any number other than one may be done either by the addition or the multiplication of the unit area. The starting point is always the unit integer. But there is no such strictly defined starting point when one defines non-quantitative words or concepts.

There one has to superimpose partially several non-quantitative concepts, each corresponding to a given area which is not strictly circumscribed. Therefore that superimposition is never absolutely fixed or definite, partly because the words constituting the definition do not have distinct contours. Any definition of a non-quantitative concept may therefore best be compared to the partial superimposition on one another of, say, six or seven patches of clouds whose contours vanish at their presumed boundaries. The meaning of the concept then corresponds to the area where all the concepts used in the definition overlap. But the area of that overlap is not strictly definite.

When viewed from a distance, such an overlap, like any cloud, appears with distinct edges. This is why non-quantitative concepts function well in ordinary discourse. But when they are subject to a close analysis, they seem to evanesce. Such is the reason for the intellectual malaise created by logical positivism. On more than one occasion it prompted despair about the possibility of knowing anything at all. One wonders whether A. J. Ayer knew the true reason why he had to admit that almost everything was wrong with logical positivism.[33] The reason was the presumption that every good reasoning should be a "scientific" reasoning. In other words, logical positivists looked for strictly defined conceptual building blocks even within the non-quantitative realm, although there only patches of fog could be found.

Yet those non-quantitative concepts do not become less real, just because it is not possible to ascribe them quantitatively exact contours. Patches of fog are just as real whether looked at from a distance or from close range. Thus the notion of forest does not become invalid just because a forest, when looked at close range, merely shows single trees. Nor does the notion of forest become invalid just because it is not possible to define quantitatively the number of trees that would constitute not merely a grove but a forest. It is not possible to find the number of pages that would necessarily constitute a book and not a mere pamphlet. It is quite an arbitrary matter when librarians, in their despair, decide that 60 pages are needed at the minimum to make a book.

While no superimposition of patches of clouds would turn them into the kind of building blocks which quantities are, quantities would not remain quantities once deprived of their strict conceptual contours. This radical difference between quantities and everything else is still to be perceived in its true weight by champions of artificial intelligence, these latter-day protagonists of reductionism. While quantitative concepts can be given "exact" equivalents in the magnetic orientation of ferro-silicate domains within the chips, this cannot be done with non-quantitative concepts that represent the overwhelming proportion of human conceptualizations. They include the crucially important value judgments and existence judgments.

Whereas this basic difference between quantitative and non-quantitative concepts may appear madness from the viewpoint of artificial intelligence, the human mind has no problem in living with that difference. It is an often underestimated marvel of the human mind that it can understand with equal ease quantities and qualities, an ease incomprehensible within the perspective of artificial intelligence programming. The human mind can grasp in a single act of knowledge entities, for instance, an aesthetically valuable painting, that convey ideas both quantitative and aesthetical at the same time and in the same act of perception. The mind is not disturbed by the fact that a human action, such as a step forward, is fully describable in quantitative terms, and yet non-describable in those terms insofar as that step was made freely.

In order to awaken the minds of my readers and audiences to this fundamental fact of irreducibility, I used to present them with a paradoxical—but I hope not irreverent—twist to a statement in the Gospel. The statement is well known: What God has joined together, no man should separate. Twisted, the phrase would go: What God has separated, no man should try to fuse together, lest confusion should arise. Human knowledge, whether we consider it to have come from the hands of God or not, concerns two separate realms, quantities and non-quantities, and these two realms are irreducible to one another. It is not profitable for man to chafe under that restriction. Those who did, whether on the Hegelian right or the Hegelian left, created only confusion for themselves and others.

About quantities, insofar as they are embodied in matter and drawn out of it by measurements and mathematical operations, science alone is competent. In that sense, and in that sense alone, science is unlimited, while remaining limited to quantities. All other considerations that relate to non-quantitative features, are beyond the quantitative competence of science which is its sole competence. Conversely, quantitative considerations, insofar as they are to be empirically verified or measured, are beyond the competence of philosophy or theology, to mention only the principal fields of inquiry that do not aim at measuring anything in sensible matter.

This distinctness between the quantitative and non-quantitative (qualitative) realms of knowledge is not proposed as a starting point in knowledge. Sensory knowledge begins with the registering of external reality, or things in short. This is true even though what is most directly perceived in things is their size. This is why the category of quantities holds first place among all categories. Or as Aquinas states, "accidents befall substance in a definite order. Quantity comes to it first, then quality, after that passivities and actions." To continue with Aquinas, "sensible qualities cannot be understood unless quantity is presupposed . . . and neither can we understand something to be the subject of motion unless we understand it to possess quantity."[34]

This primary position of quantities among all categories is the reason for their conceptual isolation among them. Quantities do not admit analogical degrees of understanding. This constitutes their radical difference from other categories and even from substance and existence. Herein lies the error of those who, with Heisenberg in the van, tried to see in wave mechanics something analogous to the Aristotelian doctrine of potency.[35] One should therefore take it for distinct progress that physics has ceased to be called natural philosophy.

The inseparability of quantities from matter justifies the quantitative character of the scientific method. Compared with it, all other considerations about science are of secondary importance, no matter how intriguing they may be. Unfortunately, for the past thirty years or so, interpretations of science have been dominated by these secondary considerations. We have learned a great deal about the psychological aspects of scientific discoveries. We have learned much about sociological factors that promote or hinder scientific progress. We have learned a great deal about paradigm shifts, research programs, scientific styles, and so forth. But because the basic feature as outlined above has not been kept in focus, a great deal of confusion has arisen about science. One result of that confusion is the view that Taoist meditation is the chief propellant of the great insights of modern physics. To see that confusion for what it is, it is enough to contrast the definiteness of numbers (including quantizations of the energy levels in the Bohr atom) with the indefiniteness of both Yin and Yang in their mutual interactions.

To cut through that confusion one need not be a scientist, one need not even be a philosopher of science. One need only to remember the role of quantities in science. It takes no advanced mathematics to ask about any proposition, whether it includes quantitatively determinable parameters. Inasmuch as it does, it is a scientific proposition. There science, insofar as it verifies or disproves theories in their quantitative inferences, alone is competent. Insofar as that proposition contains parameters other than quantitative, other branches of human discourse—philosophy, theology, esthetics, or whatnot—should be resorted to in order to evaluate them.

This multiple approach demands mental discipline in no small measure. First of all, the scientist should be aware of the fact that even the most appealing procedure may not be free of ambiguities. Symmetry in equations may seem to recommend itself on aesthetic grounds as an unquestionably worthy and fruitful goal. Yet those grounds will appear somewhat shaky as soon as one considers that there is nothing symmetrical in the "golden cut" or golden proportion in which ancient and modern artists have recognized something profoundly aesthetic. Also, an absolutely perfect symmetry is not applicable in relation to a physical universe in which nothing would move if there were not some basic imbalance built into it.

To overlook such ambiguities will not come easily to those whose chief training is in the one-way thinking of quantitative method and in nothing more complicated. They will be swayed time and again by the staggering measure to which matter can be manipulated through its quantitative properties. Science extends to wherever quantitative properties can be found and is competent to handle them. Beyond that, science is not only incompetent, but may be the source of most dangerous expectations. I have often stated that over the entrance of every laboratory and department of science one should carve the words of Maxwell, which I quoted above, about the severest test of the scientific mind. To those words should be joined a warning by Polykarp Kusch, a Nobel-laureate physicist. "Science," he said, "cannot do a very large number of things, and to assume that science may find a technical solution to all problems is the road to disaster."[36] To safeguard against such a disaster few considerations may be more effective in this scientific age than a reflection on the limits of an otherwise limitless science.

[1] As told by C. P. Snow, "The Moral Un-neutrality of Science," *Science* 133 (1961), p. 257.

[2] Lord Kelvin, "Electrical Units of Measurements" (1883), in *Popular Lectures and Addresses* (London: Macmillan, 1891-94), vol. I, pp. 72-73.

[3] R. Mayer, "Ueber notwendige Konsequenzen und Inkonsequenzen der Wärmemechanik" (1869) in *Gesammelte Schriften* (Stuttgart, 1893), p. 355.

[4] Letter to Greisinger, July 20, 1844, in *Die Mechanik der Wärme* in *Gesammelten Schriften* (Stuttgart, 1893), p. 145.

[5] A. Einstein, "What is the Theory of Relativity?" in *The World as I See It* (New York: Covici, 1934), p. 80.

[6] See E. Warburg, "Friedrich Kohlrausch: Gedächtnisrede," in *Verhandlungen der Deutschen Physikalischen Gesellschaft* 12 (1910), p. 920.

[7] For details, see my *The Relevance of Physics* (Chicago: University of Chicago Press, 1966), pp. 127-30.

[8] See E. Fuller (ed.), *2500 Anecdotes for All Occasions* (New York: Avenel Books, 1978), p. 192.

[9] H. Hertz, *Electric Waves*, tr. D. E. Jones (London: 1893), p. 20.

[10] R. Feynman, R. B. Leighton, M. Sands, *The Feynman Lectures on Physics* (Reading, MA: Addison-Wesley, 1964), vol. II, section 41, p. 12.

[11] Ibid.

[12] See my review, "Mind: Its Physics or Physiognomy?" (1991) of Penrose's book, reprinted in my *Patterns or Principles and Other Essays* (Bryn Mawr, PA: Intercollegiate Studies Institute, 1995), pp. 204-213.

[13] As I argued in my review of Hawking's book, "Evicting the Creator" (1988), reprinted in my *The Only Chaos and Other Essays* (Lanham, MD: University Press of America; Bryn Mawr, PA: Intercollegiate Studies Institute, 1990), pp. 152-161.

[14] See my *God and the Cosmologists* (Edinburgh: Scottish Academic Press, 1989), pp. 138 and 258. New enlarged edition, Real View Books (Fraser, MI, 1998).

[15] D. Hume, *An Enquiry concerning Human Understanding*, Section 12, part 3, in *Philosophical Works* (1826), vol. 4, p. 193.

[16] In a conversation with Carnap. See the latter's "Intellectual Autobiography," in P. A. Schilpp (ed.), *The Philosophy of Rudolf Carnap* (La Salle, IL: The Library of Living Philosophers, 1963), pp. 37-38.

[17] Just as Einstein viewed, in the name of science, free will as illusion, so did he view any abiding sense of human purpose and, of course, the immortality of the soul. For details and documentation, see my *The Purpose of it All* (Lanham, Md: Regnery Gateway, 1990), pp. 182-83.

[18] "C'est librement qu'on est déterministe," goes Poincaré's priceless phrase in his essay, "Sur la valeur objective des théories physiques," *Revue de métaphysique et de morale* 10 (1902), p. 188.

[19] Reported in *Trenton Times Advertiser*, February 25, 1995, p. 1.

[20] Reported in E. Gilson, *From Aristotle to Darwin and Back Again. A Journey in Final Causality, Species, and Evolution* tr. J. Lyon, with an introduction by S. L. Jaki (Notre Dame, IN: University of Notre Dame Press, 1984), p. 28.

[21] A point wholly ignored in standard high-level popularizations of brain-mind research, such as D. J. Chalmers, "The Puzzle of Conscious Experience," in *Scientific American* December 1995, pp. 80-87.

[22] W. Quine, "Ontology and Ideology Revisited," *The Journal of Philosophy* 80 (1983), p. 1.

[23] In numerical terms the efficiency in question is 16%.

[24] See *The Scientific Papers of James Clerk Maxwell*, ed. W. D. Niven (Cambridge, 1890), vol. II, p. 759.

[25] See chs. 5 and 6 in my book, *Is There a Universe?* (Liverpool: Liverpool University Press, 1993).

[26] *Categories*, 6a. I am quoting the English translation in *Great Books of the Western World*.

[27] Ibid.

[28] Those who, in imitation of A. Koyré, contrast in such a way pre-modern with modern science, usually fail to refer to that phrase of Aristotle who himself overlooked its import in dealing with the physical universe.

[29] For a list of plain absurdities in Aristotle's physics, see my *The Relevance of Physics*, pp. 26-28.

[30] This puzzlement is the gist of the exchange of letters between Pierre Duhem and R. Garrigou-Lagrange. See my article, "Le physicien et le métaphysicien. La correspondance entre Pierre Duhem et Réginald Garrigou-Lagrange," *Actes de l'Académie Nationale des Science, Belles-Lettres et Arts de Bordeaux* 12 (1987), pp. 93-116.

[31] See my paper, "Words: Blocks, Amoebas, or Patches of Fog? Some Basic Problems of Fuzzy Logic, in *Proceedings of the International Meeting of the Society of Photo-Optical Instrumentation Engineers, Orlando, Florida, April 10-12, 1996,* ed. B. Bosacchi and J. C. Bezdek (Bellingham WA: SPIE, 1996), pp. 138-143.

[32] See A. N. Whitehead, *Modes of Thought* (1938; New York: Capricorn Books, 1958), p. 235.

[33] Quoted in B. Magee, *Men of Ideas: Some Creators of Contemporary Philosophy* (London: British Broadcasting Corporation, 1978), p. 131.

[34] St. Thomas Aquinas, *The Division and Methods of the Sciences. Questions V and VI of his Commentary on the De Trinitate of Boethius.* Translated, with Introduction and Notes by A. Maurer (4th rev. ed.; Toronto: Pontifical Institute of Medieval Studies, 1986), p. 40 (Qu. V. art. iii).

[35] For details, see my *God and the Cosmologists*, pp. 155-56.

[36] Address to the Pulitzer Prize jurors, Columbia University, 1961, in *New York Herald Tribune*, April 2, 1961, sec. 2, p. 3. col. 5.

2

Extraterrestrials,
or Better Be Moonstruck?

"Are we alone? . . . Statistically, there is every likelihood that life has evolved elsewhere in the universe." So it was claimed in the "Millennium Notebook" about "questions that stump scientists," in *Newsweek*'s January 19 [1998] issue.

When someone proposes the probability of the origin of life in the context of the question, Are we alone? one may shift the issue directly to another question: What is the probability that there are highly developed technological civilizations elsewhere in the universe?

This probability is far from being as favorable to the search for extraterrestrials as is it generally believed. Yet the media takes lightly, or simply ignores, eminent scientists who have expressed scorn for the idea that there are extraterrestrials able to communicate with us. Ernst Mayr, the dean of American biologists, said the federal funding of SETI (Search for Extraterrestrial Intelligence) was a "deplorable waste of taxpayers' money." He had no choice. As a

---

An enlarged form of an article that first appeared in *National Catholic Register*, Feb. 15-21, 1998. Reprinted with permission.

consistent Darwinist he had to regard the emergence and further evolution of life as a chance process. Therefore he had to view it as most improbable that evolution would repeat itself elsewhere and produce intelligent beings similar to us.

Such statistical considerations were in the mind of Enrico Fermi, who achieved the first controlled nuclear fusion in 1942, as he dismissed the idea of visitors from outer space with the remark: "If they exist, they would have long ago landed on the lawn of the White House." They would have certainly done what is far easier, namely, awakened us up with their radio signals and removed our cataracts with their powerful laser rays.

On a purely Darwinian basis, the Nobel laureate physicist C. N. Yang hit the nail on the head when thirty or so years ago he suggested that we must not try to answer any radio signal from another civilization. No other attitude is reasonable from the viewpoint of Darwinian theory which offers no exception from a remorseless struggle for life with no quarters given. It is only in Isaiah's eschatological vision that a lamb lies down with a lion and a child plays with a viper.

One need not be an expert in the life sciences or in nuclear physics to realize that instead of "every likelihood" one should talk of an improbability of well-nigh zero. It is enough to ponder the presence of the moon around the earth to be struck by that improbability. This presence is unique in the solar system, although there are scores of moons around the other planets. The mass of the moon relative to the earth, its chemical composition remarkably similar to the chemistry of the earth's mantle, the moon's apparent size, its daily and monthly influence that produces the tides—all these make the earth-moon system unique in the solar system. It is sheer science to say that we earthlings live not simply on the earth but on the earth-moon system.

The earth as well as the other planets may have originated in a rotating sun. Again, many moons of other planets may have perhaps originated through the rotation of their respective planets. In that case they may be considered as more or less typical occurrences. But the origin of our moon cannot be explained in this "likely" way.

One of Darwin's sons, George Howard Darwin (1845-1912), an astronomer, showed that the moon gradually recedes from the earth, and therefore, hundreds of millions of years ago must have been very close to it. He could have shown right there and then that the moon had to originate from the mantle of the earth.

Today astronomers dealing with the origin of the moon accept the unlikely scenario of a glancing collision between the earth and a hypothetical celestial body called X. This scenario contains at least five independent factors, all rather unlikely. The body X had to have a mass ten times the mass of Mars (1). The direction (2), the velocity (3), and the plane (4) of the motion of X had to be within very narrow margins so that our moon and our earth-moon system might be the result. Moreover, the collision had to occur within a narrow period of the formation of the Earth itself (5)!

If one now takes the probability of each of those factors for one in ten, or 10^{-1}, which is a most conservative estimate, their combined probability is one hundred thousandth, or 10^{-5}. Actually, it would be more accurate to say one in a million, or 10^{-6}, because factor 5 can hardly be given a greater probability than one in a hundred. The figure 10^{-6} would alone undermine the likelihood that it is reasonable to look for radio messages sent out from other technological civilizations towards the earth.

Ten thousand, or 10^4, is the typical figure given by supporters of SETI as the number of technological civilizations in our galaxy. This is the figure which Frank Drake supports in his latest evaluation of the Drake equation which he first proposed in 1961. This figure is based on taking the number of stars similar to our sun in our galaxy. Obviously only a fraction of such stars would have a planetary system around them. Only a fraction of such systems would have an earthlike planet. Only on some of such planets would life evolve and evolve in turn into higher organisms. And only a few types of these would reach high intellectual and technological levels.

So much for the way Drake and others have reduced the figure ten billion (the number of stars in our galaxy) to a mere ten thousand. But they have invariably disregarded the unlikelihood of the earth-moon system. Had they done so, they would have arrived at 10^{-2}, the product of 10^4 and 10^{-6}. This would mean that the likelihood of there being a single technological civilization other than ours in our galaxy is one in a hundred. This in itself would be a far cry from "every likelihood," even if one does not take a lower improbability for all those five factors. Communication with civilizations of extraterrestrials in other galaxies, let alone a visit from them, certainly belongs to the cover of cereal boxes, where E. Purcell, a Nobel-laureate physicist, put the idea of interstellar travel almost four decades ago.

The moon played a crucial role not only in the evolution of purely organic life by producing the tidal basins, but also in the development of man's intellectual life. Suffice it to recall Aristarchus' measurement of the relative and absolute distances among the earth, the moon, and the sun. Without this measurement there would not have been a Ptolemaic astronomy. Without Ptolemaic astronomy there would have been no Copernican astronomy, and without Copernicus no Newton.

Yet Aristarchus' feat would not have been possible if the moon's apparent diameter in historic times had not been equal to that of the sun. Without the moon being where it is—no nearer and no farther from the earth—Newton could not have convinced himself that the celestial bodies obeyed the same laws of motion as did the freely falling bodies on earth. In view of this, the accidental fall of an apple on young Newton's head takes on a new significance.

In other words, before one waxes enthusiastic about extraterrestrials, one had better be ready to be a bit struck by what the moon means to the earth. This would have been, of course, the duty of astronomers like Drake and others who are spreading the gospel of "every likelihood." They assume without further ado that once there is life, there is intelligence, and once there is intelligence, there is science and advanced technology.

The history of science shows exactly the opposite. Science suffered a monumental stillbirth in all great ancient cultures, such as China, India, Egypt, Babylon—and Greece as well. None of them turned out to be the matrix for the formulation of Newton's three laws, the very foundation of exact science and technology.

Of those three laws Newton formulated only the third, the force law. The second law (action equals reaction) he borrowed from Descartes. The first, the most fundamental, the law of inertial motion, was formulated by John Buridan, more than three hundred years before Newton. And he formulated it in the context of his Christian belief of creation out of nothing and in time.

Christian faith, a unique reality on earth, is, of course, inconceivable without the Incarnation, another unique event. Let us consider here improbabilities unrelated to religion. Buridan might have perished in the Black Death of 1349 that claimed one third of Europe's population and ravaged Paris too. There would have been no Kepler's laws, the very foundation of Newtonian physics, if Tycho Brahe had lost not only his nose in a duel, but also his very eyes.

There would have been no Newtonian system, if young Horrocks, the author of the first readable account of Kepler's laws, had died not at the age of 21 but at the age of 18. Geniuses, let us not forget, cannot be had on order, like so many take-out lunches.

So much for some very narrow escapes for science, which had many more such escapes as can be seen by any non-triumphalistic account of the history of science. Their combined improbability might easily reduce the one-hundreth probability to one millionth, or to perhaps a billionth or even less. In other words, instead of talking blithely of "every likelihood," one might say that the probability of finding at least one group of technologically accomplished extraterrestrials in our galaxy is utterly minimal on the basis of what we know, rather than what we may imagine in brazen disregard of facts.

But this is not yet the whole of the advisability of letting oneself be a bit moonstruck first, before speculating about extraterrestrials. The moon, as anyone can find out with good binoculars, is pockmarked all over. Even more so on its far side. There one of the largest craters—about 12 miles in diameter—is called Giordano Bruno. It now seems certain that it was caused by the impact of a huge comet or meteor.

Furthermore it is possible to date that event with fair certainty. It is known that the longitudinal free librations of the moon are slowing down (being dampened). Since they could not have a starting magnitude above a certain maximum value, the past duration of those librations can be estimated. This work was done by the astronomer J. B. Hartung in 1976, and further refined by O. Calame and J. D. Mulholland, who utilized the data obtained by the Luna 24 mission and by laser range observation. Their conclusion was that those librations could not have started much earlier than about eight hundred years ago. They also pointed out that the impact of a huge meteor just beyond the edge of the moon must have started those librations and that the fiery explosion produced by the impact might have been seen from the earth.

Somehow those two astronomers learned about a strange detail in the famed *Chronicles* of Gervase of Canterbury, concerning the night of June 18, 1178. They quickly saw that 800 years lead one back more or less to that year. On that night, Gervase (the best medieval chronicler of England) and at least five others saw that

> the upper horn of the New Moon suddenly split in two and from the midpoint of the division a flaming torch sprang up, spewing out fire, hot coals and sparks to a considerable distance. Meanwhile the rest of the

moon's body became, so to speak, anxiously twisted, and convulsed as if it were a snake, to use the words of those who reported this to me as something which they saw with their own eyes. After that the moon returned to its normal state. This vicissitude was shown by the moon more than a dozen times, namely, that it sustained, as if drunken, various fiery torments and returned again to its normal shape. After these vicissitudes the moon became sort of blackish from horn to horn. These things, which I am writing, those men, who saw them with their very eyes, were ready to confirm under oath, namely, that they added nothing false to the details given above.[1]

But suppose that the comet had arrived a bit later and instead of hitting the moon it had crashed into the earth. Had it hit the earth somewhere in Western Europe, it would have extinguished the nascent university system and would have snuffed out the very medieval beginnings of modern science.

Only those who are able to see the hand of Providence behind the moon's posing as a shield for the earth have nothing to fear. The immensity of outer space opening up in the 17th century frightened, as Pascal well put it, only the libertines, the "freethinkers" of his time. The terribly catastrophic character of cosmic spaces, as it is coming into view today, should seem hopelessly terrifying only for those who have nothing to see beyond those cosmic vistas. Today they dream about extraterrestrials, because they are afraid to be alone. Rather they should be ready to be a bit moonstruck, and to do so in the name not so much of religion, but of plain science. They might then even notice that beyond science revealed religion looms large as its saving grace.

---

[1] See the 1879 English edition, vol. 1, p. 276.

3

Computers: Lovable but Unloving

Modern society craves for love and all too often fails to find it. So many illustrations are the various ways in which love is celebrated in modern life. Divorces have become occasions for formal celebration. Nothing specific should be said of the spreading of pathetic forms of love, such as one-parent families, or of some repulsive forms, such as megafamilies, that could be called the megabytes of sex. All too many forms of celebrating love have become trivial matters, such as children's birthdays. They are less about children's love for one another, which should not be overidealized, than about children's insatiable appetite for gifts, mostly in forms of toys. Toys in turn are discarded at the first moment when a new toy appears on the scene.

The fact that children easily become tired of everything, including toys, inspired the observation about the world, that it is a big child and therefore tires easily. This is almost literally true of the world of computers, where PCs (personal computers) have

Invited lecture for the IUVE conference in Madrid (1993). Published in *Downside Review* 112 (July 1994), pp. 185-200. Reprinted with permission.

replaced mainframe computers as prime merchandise. PCs came to the scene in 1981, when IBM introduced its first model. The anniversary was marked with big headlines in July 1991 together with a reference to the fact that the number of PCs had grown in a decade to over 60 million. (With the introduction of notebook computers, the growth has become even faster). On reading those headlines not a few PC owners and users, whose total number now may be close to half a billion, may have been prompted to doodle a "Happy Birthday to you, dear PC," and especially if they worked for IBM, Apple, and Toshiba. For those companies, and for many others, PCs are certainly a lovable thing in that sense in which love sparks a desire for exclusive possession.

That PCs are lovable things is also shown by the fact that demand for them is insatiable. And just as ladies who never fail to show up in something novel, computers come out in new models at an accelerated rate. By the time the latest model is available, a newer model is already on the drawing board. The increasingly smaller models of computers also indicate that they are lovable creatures. That laptops rhyme with lapdogs, these quaint objects of love, has indeed been noted. Love has always preferred to see its objects in the perspective of diminutives. PC's have already generated a market for a large number of accessories which come packaged with a touch of glamor. Last but not least, computers can trigger what has been a perennial feature of love, namely, its ability to provoke hatred. This love-hate syndrome may have been at play when the frustrated owner of a PC reached out for his shotgun and fired a round into his PC after it refused to "communicate" with him. PCs are apt to refuse to enter into a symbiosis with their owners, to recall an expression from a new language which is generated by computer technology and programming. Here too as elsewhere, love creates its own dictionary which is understood only by members of the club of the initiated, if not infatuated.

As all beginner users of computers have to learn at their own cost, computers co-operate only when given accurate, error-free commands. This alone shows that they have no intelligence in the human sense, the only sense in which intelligence can be understood. To realize that one has made an error is the privilege of the human intellect. There is an unsuspected philosophical depth in the old Latin phrase, *errare humanum est*. The phrase reflects the wisdom of such a giant of Classical Antiquity as Aristotle who was not reluctant to state that man passes much of his life in error.

Thomas Aquinas said much the same for philosophical as well as for theological reasons. Hylemorphism, when applied to human nature which is a union of body and soul, implies the possibility of many errors. Error becomes a problem only for materialists. It is they who have to conclude that "all the errors of man are errors made in physics."[1] So wrote Baron d'Holbach, a leader of the outspokenly materialist wing of the French Enlightenment. What he failed to consider was the problem of how there could be an erroneous physics in the first place, whatever the endless parade of mistaken physicists.

The good Baron would have found the answer had he not attributed, as did many of his allies, patently absurd views to Thomas Aquinas. Only one of them, Diderot, was willing to admit that he had not read the works of the Angelic Doctor. They could have learned from his works that error is possible only if, in addition to matter, there is reason, confined to that bulk which is called body. In a world of rapidly changing empirical conditions, it is not reasonable to expect that the mind should have for its disposal all the sensory data all the time in order to make right judgments at any moment. The errors that can be made by the mind should seem even more numerous in a world of fallen nature where the sensory data supplied to man's appetites are not necessarily under man's rational control.

Such a view, a very realistic and sane view of the human intellect, could still be voiced at the time when the first calculating machine was being constructed, in the middle of the 17th century. The famed inventor was Pascal, a man deeply conscious of man's fallen nature. Yet the fallenness of man is not the sole justification of Pascal's remark that "the most powerful cause of error is the war existing between the senses and the reason."[2] Pascal saw that not all sensory impressions, so basic for the formation of concepts and ideas, are available to man all the time and in the right order. Because of equally basic considerations about the immense surplus involved in any thought as contrasted with mere sensory impressions, Pascal was not to attribute thinking to his calculating machine: "From all bodies together, we cannot obtain one little thought; this is impossible, and of another order."[3] So spoke the author of the famed *Pensées*, who also had a warning for those who took the calculating machine for much more than it was. According to him, the calculating machine was a mere time-saving tool that made it possible for man to go through the drudgery of computing "without having his mind on the job."[4]

Apart from that invariable inequality between sensory perceptions and thoughts, Pascal had another, and even more decisive, reason to deny intelligence to machines. He saw something enormously revealing in the fact that man always reasons for a purpose. This means that the will plays an integral part in every human deliberation, however trivial, let alone when that deliberation is a lengthy and complicated process. (As will be seen later, the role of the will is, has always been, and shall forever remain crucial in the pursuit of every scientific truth.) While the calculating machine could create the illusion that it embodied some intellect, no illusion could be entertained with respect to a calculating machine with a will. Pascal was peremptory: "The arithmetical machine produces effects which approach nearer to thought than all the actions of animals. But it does nothing which would enable us to attribute will to it, as to the animals."[5] In other words, Pascal emphatically denied to calculating machines even an appetitive will, let alone a will which is genuinely free and therefore can choose against any and all appetites.

It would have taken Pascal's intuitive powers to see a further evidence of that absurd war between man's senses and reason when a decade or so after Pascal's death, Leibniz proposed the first modern computational theory of the mind. I mean his universal calculus, or his planned reduction of all mental operations to a system of binary counting, the very basis of all computers, personal and other. That a great mind, like Leibniz, could be so shallow with respect to what is understanding, would have been taken by Pascal for an example of that absurd war. That war did not end when Pascal's calculating machine was developed by Leibniz to the point where it could perform multiplication by rapidly done additions.

What Pascal would have found most deplorable in Leibniz's infatuation with mathematics concerned the will. The will always acts for a purpose, or else it is not a will, but a blind passion. It is that will, free and purposive, which Leibniz hoped to justify with mathematical physics. He did not see that he would commit thereby that elementary error which is to put the cart before the horse. Leibniz's hope made no sense, unless it was a purposive hope which had to exist in its own terms before it could be cast into a mathematical formalism. But this is what Leibniz failed to perceive as he urged, on several occasions, that man's conviction of acting for a purpose would not become a well-demonstrated proposition until it had been given a mathematical form.

What Leibniz had particularly in mind related to final causes that are at work in every purposeful action. Not that his intentions were not good. He wanted to neutralize the threat posed by the new mechanistic science to man's appreciation of his own purposive action. It should be enough to think of Francis Bacon's diatribes against final causes as having no place in science and, by implication, in respectable human reasoning. But in Leibniz's case too, good intentions proved to be a road to intellectual perdition. Final causes could not be saved if it were true that, as Leibniz wrote, "Very far from excluding final causes and the consideration of a being acting with wisdom, we must from these deduce everything in Physics."[6]

This was, of course, bad enough for physics, but even worse for final causes. For what Leibniz really meant was that physics had to justify any discourse about final causes, or at least that we must have a form of physics that readily accomodates all such discourse. That such was in fact Leibniz's idea is very clear from his reference to "what Socrates in Plato's *Phaedo* admirably well observed in arguing against Anaxagoras and other philosophers who were materialistic."[7]

What Socrates really argued was a new way of proving the immortality of the soul, that very entity which was driven by such final causes as the ideal of absolutely valid moral norms—truthfulness, for instance. The particular issue on hand was whether Socrates would act morally if he were to escape from jail and give thereby the impression that he had a guilty conscience. To Socrates' friends, who were under the influence of Anaxagoras' materialistic physics, it made no difference whether one died with a guilty conscience or not. For them everything was matter, fully understandable by mechanistic physics, beyond which nothing else was to be understood. They could not be impressed by metaphysical proofs of the immortality of the soul. Socrates therefore decided that it was necessary to give a physical demonstration of the soul's existence. He thought that this was possible if he first proved that physical bodies themselves acted for a purpose as they moved. If one accepted that matter was goal-seeking in the purposive sense, it was unreasonable, Socrates argued, not to consider spiritual or moral purposes as being so many pointers to a non-material state of affairs, identified with the immortality of the soul.

Leibniz failed to see that in reasoning in such a manner Socrates threw out the baby (mechanistic physics—a very good physics, however incomplete) with the dirty bathwater which is mechanistic philosophy.[8] Yet there was nothing wrong in what introduced that reasoning, a reasoning so fateful for the future of science. Fateful, because it became fully systematized in Aristotle's purposive physics, a physics very distinct from his philosophy of physical nature. That physics has few correct propositions in it.

To prove the immortality of his soul and therefore justify his readiness to drink the hemlock, Socrates began with arguments that are classic even today. Such an argument is, for instance, the ability of man to generalize, an ability at work in the formation of each and every human word. Every human word is a universal. It is on that basis that Plato postulated a universe of ideas, a universe free of the constraint of matter and of changes, those invariable concomitants of matter. The soul was to return to that realm of immortality precisely because it could not help finding universal truths in material situations that are always particular. Plato would have found very much to his liking Pascal's dictum about the absurd war between man's reason and his senses.

Warring or not warring, the mind (or soul) and the human body have very different activities. Consideration of those differences has always constituted a chief reason for concluding that the soul has to be immaterial and therefore immortal. In a context about computers it should be enough to think of such products of man's mind as, for instance, the notion of zero, or the notion of imaginary and irrational numbers, or the difference between summation through finite series and the going to the limit through infinitesimals.

Infinitesimals have served us so well in practice as to make us insensitive to some remarks about them when they were novel and therefore without that protective garb which custom and familiarity provide. One such remark, made by Bishop Berkeley two hundred and fifty years ago, is worth recalling. He ridiculed fluxions as being neither a finite quantity, nor a quantity infinitely small, nor yet nothing. He called fluxions, or instantaneous rates of change, "the ghosts of departed quantities" and reminded "the philomathematical infidels" that those "who can digest a second or third fluxion . . . need not, methinks, be squeamish about any point in Divinity."[9]

Champions of artificial intelligence must indeed have chronic mental indigestion. This sickness hits them through their insatiable appetite to reduce the mind to mathematics. One need not even bring up Gödel's theorem of incompleteness to show them that the mind is far more complete than all mathematical theorems taken together.

Much less should Gödel's theorems be brought up if this is done in a way which is equivalent to starting a march through the vain effort of first making the second step. For the principal point in Gödel's theorem, a point unrealized by Gödel himself, is not that formalization cannot form a self-consistent system and that regress to infinity is but an endless postponing the answer. The principal point lies in the tacit assumption that formalization, mathematical or other, is a form of consistent thinking. In other words, it does not make sense to talk about the incompleteness of mathematics unless one grants in the first place that there is mathematics or consistent quantitative talk. Only the incompleteness of mathematics can be demonstrated through Gödel's theorem but not its very existence. Or in still other words, any analysis of formalization is possible only if one first admits there are forms, mathematical or other.[10]

If one forgets this, one has no choice but to turn Gödel's theorem into the quizzical or quixotic proposition that all formalization is incomplete and this is the only complete truth. To see the true weight of Gödel's theorem, one must muster more intellectual honesty and courage than that evident in Auguste Comte's coping with the inner contradictions of his positivism. In the latter everything was relative, according to Comte's own admission, because positivism meant the overcoming of the mythical and metaphysical phases of human thought, both of which were dominated by absolutist ideas. "Everything is relative and this is the only absolute truth," was all that Comte cared to say in answer to objections that he contradicted himself in claiming a final or absolute character for his system of positivism.

The limits of formalization constitute the kind of objections that should seem to be most appropriate when facing advocates of artificial intelligence. The most obvious examples of those limits are on hand in every entry in every dictionary of every language. To define the meaning of a word, one has to rely on the meaning of other words by making those meanings partially overlap. There can be no total overlapping because in that case the word, whose meaning is to be defined, would mean exactly the same as do the words used to define its very meaning. The overlapping, which has to be partial, has, however, a very special feature. It can best be illustrated by partially superimposed circles or other geometrical figures that themselves do not have an exact contour. This is why language translation by computers will forever remain a most imperfect technique which shall demand human interpreters to make it work, if it will ever work at all, even at a moderately acceptable level.

In other words, nothing is certain about words in the sense of arithmetic certainty. The word "certain" is the most tangible proof of this. In fact, it was no less a physicist than Bohr who as a philosopher tried to make the most of the strange suppleness of the word "certain." There is hardly a paragraph in his philosophical writings where he would not use the expression "in a certain sense," or "to a certain degree." In each case he meant exactly the opposite, namely, that the idea he tried to convey could not be given a definition which is certain in the sense of being strictly defined or definable.

Unfortunately, Bohr the philosopher, and he was a shabby one, relied heavily on those expressions in order to shield himself against strict counter-arguments.[11] Even more unfortunately, most of his readers readily complied with this tactic of his. In this age of computers and PCs, imprecise thinking paradoxically has become the standard of scholarship, or rather of a scholarship suspicious of standards or norms. To speak of a standard, or a sort of a yardstick, and actually to mean exactly the opposite shows, however, the main point to be made in this essay. Language consists not so much of words as of phrases, or contexts that are not only immensely more numerous than words, but also unpredictable. Herein lies the source of innovative style.

A most important fact about words is that they are arbitrary signs. This is why there are thousands of languages with an incredibly wide variety of vocal signs or phonemes to denote one and the same thing or idea. Even the word denoting the being called mother in English need not necessarily begin in all languages with the soft sound m indicative of a mother's usual kindness. Words are therefore so many acts of the will, arbitrary as that will may appear in forming words. The same arbitrary will is at play in the forming of all diacritical signs, such as the comma, period, colon, semicolon, question mark and so forth. In all modern languages it is a period (a dot or point) that indicates the end of a sentence. Would it not have been more logical to use a short vertical line?

This question may lead us to the question mark. It would not be a bad idea if all languages would adopt the Spanish usage, in which a question is preceded with a question mark and not only followed by one. Such a mark at the very head of the phrase would make it immediately clear that a question is on hand. But it would still remain purely arbitrary to write that question mark upside down. No sign is really logical, not even the sign of exclamation! This is also true of all signs used in that apparently most logical of all fields, mathematics. It should seem most arbitrary, that is willful, to use + for addition instead of — as the latter would be more symbolic of joining two entities than the former in which the vertical could easily be taken for a canceling out of that process. In particular, the sign of zero should appear a most arbitrary sign. The mathematical symbol 0 is something, and yet it is meant to stand for nothing. The symbol 0 may be the greatest evidence of the free creativity of the human intellect. Freedom is, however, a characteristic of the will. To be free to think or not to think is not primarily an act of the intellect.

What is the logic in the sign of square root or even in the symbol for infinity? Would not a circle have done much better? Less arbitrary may seem the sign of the swastika, the symbol of eternity in many cultures. Most illogical and therefore arbitrary should seem, however, the fact that the verbal form of that sign in sanskrit means well being. For nothing can be more repulsive than, say, the prospect of reading the same book over and over again an infinite number of times.

Science did not fare well at all in ancient cultures that were invariably dominated by the idea of eternal recurrences.[12] This idea numbed the will to look for discoveries. Insofar as science is a chain of discoveries, it is an enterprise of the will as much as, if not more than, of the intellect. Indeed the lives of some prominent scientists resemble a most deliberate, and at times very willful pursuit of a goal. In Galileo's very words, Copernicus was ready to commit a rape of his intellect in proposing the heliocentric system. Kepler's groping for his three laws was a total commitment, an act more volitional than intellectual. To think deeply on one and the same topic over many years implies the will as much as the mind; this was the case with the discovery of the law of gravitation, according to Newton himself.

In the early 19th century will played a crucial part in the discovery of electromagnetic induction, as can be seen in any biography of Oersted, Henry, and Faraday. Planck worked feverishly in order to become the first to discover the quantum of action. In our times, the race for the discovery of the double helix became as noteworthy as notorious. The discovery of Pluto in 1930 demanded an arduous scanning of several hundred thousand square centimeters of photographic plates over many years. The discoverer, W. C. Tombaugh, must have had a very strong will indeed. Enormous resolve propelled forward the construction, by Charles Babbage, of the first modern mechanical computer a century and a half ago. The same central role of the will is on hand in the cutthroat competition for the smallest and fastest chips and software.

Equally revealing are some telltale remarks made by those committed to the reduction of the human mind to a mere computer. Confronted with his failure to achieve his aim, Hofstadter urges the reader of his *Gödel, Escher, Bach* "to confront the apparent contradiction head-on, to savor it, to turn it over, to take it apart, to wallow in it, so that in the end the reader may gain new insights into the seemingly unbridgeable gulf between the formal and the informal, the animate and the inanimate, the flexible and the inflexible."[13] Not by the farthest stretch of the imagination can these actions—to savor, to wallow—be seen as exercises in logic. At any rate, did not Hofstadter expect his readers to engage in those actions freely?

Had he faced up to this question, he would have had to consider the problem of how to program an action, or even a mere thought for that matter, insofar as it is free. In all likelihood he would have muttered something about randomness. It is most unlikely that he would have gone any further. Randomness and chance are buzzwords which few scientists dare to look at closely. For a close look would demand a close scrutiny of the theory of random numbers and of the true meaning of Heisenberg's uncertainty principle.

As to the theory of random numbers, it provides no real randomness. It should be enough to think of a remark of John von Neumann, one of the foremost mathematicians of this century and one who contributed most to computer theory. According to him, "Anyone who considers arithmetical methods capable of producing random digits, is, of course, in a state of sin."[14] Let, therefore, those who think that they can program true randomness in terms of binary calculus, which is based on integers, also consider how to program the state of sin, let alone the state of original sin. The problem of a guilty conscience, which Socrates grappled with, cannot be exorcised with facile references to binary calculus and feedback mechanism.

As to Heisenberg's principle, it has two meanings. Taken in its good or scientific meaning, the principle states nothing more than that there are operational limits to the accuracy of measurements. But insofar as that principle has become the capsule formula of the

Copenhagen interpretation of quantum mechanics, it has taken on a meaning which has nothing to do with science and is also very shabby philosophy. Like all mistaken philosophical meanings, this too rests on a fallacy in logic. Here the fallacy is the jumping from the operational to the ontological level, a move which should be clear once the denial of causality in terms of Heisenberg's principle is rendered as follows: "An interaction that cannot be measured exactly, cannot take place exactly." The fallacy is to equivocate with the word "exactly" by not distinguishing, in one and the same inference, between its two very different meanings: One meaning is operational, the other ontological.[15] So far we are on the level of pure logic.

Beneath the logical defect of taking Heisenberg's principle for a refutation of causality, there is the emotional ground for taking it for such a refutation. It is still to be widely realized that Heisenberg had denied causality for some years before he formulated his principle in 1927. He was still in his late teens when he denied causality under the influence of the vitalism he encountered in the philosophy of the so-called Student Movement (Jugendbewegung). A few years later he found that some prominent older physicists had been denying causality for similarly non-logical reasons. Ultimately, all such reasons rested on that voluntarism which Kant had generated by his subjectivism.

This voluntarism has become a distinctive feature of modern Western culture. It is as much alive in totalitarian ideologies as it is in various forms of pragmatism that support the ideology of Western democracies. In none of them is there room for a proper respect for free will. Disrespect for free will had for some time been promoted by prominent scientists before computer theorists and champions of artificial intelligence became emboldened to ignore free will for all practical purposes. As they freely go about their implicit and at times explicit denial of freedom, they resemble the Darwinists whom Whitehead once dismissed with a devastating phrase: "Those who devote their lives to the purpose of proving that there is no purpose, constitute an interesting subject for study."[16]

The same applies to those who freely deny free will. They still have to learn the elementary fact that no materialist is reluctant to make up his mind and no determinist argues deterministically. Such are Chesterton's remarks. Poincaré voiced the same elementary truth when he remarked that "c'est librement qu'on est déterministe."[17] It makes no favorable reflection on modern society that it is ready to listen to such elementary and non-mathematical wisdom only when it comes from a prominent mathematician.

Or from a philosopher, but only when the philosopher in question is a mere *philosophe* rather than a true lover of wisdom. Diderot was one such *philosophe* who did not differ from other *philosophes* in liberally experimenting in free love. It was through his most remembered love-liaisons that he discovered the agony which is in store for all who take man for a mere machine and try to be a bit consistent about it. On being asked by his lover, Mme de Maux, whether comets blindly obey the law of gravitation, it suddenly dawned on Diderot that his love for Mme de Maux might then be just as well a blind submission to fate. "It makes me wild," he wrote to her, "to be entangled in a devil of philosophy that my mind cannot deny and my heart gives the lie to."[18]

Champions of artificial intelligence would make some progress toward sanity if they were to ponder how to program free actions, how to program love, these two crucial fulcrums of human life, including its intellectual form. This is not to suggest that they have in any way shown any promise of success in programming thought and consciousness even in their most intellectualized forms. It is quite legitimate to press them on these purely intellectual points. It is quite legitimate to confront them with the question about the difference between a sign, be it a mere word, and the idea carried by it, let alone about the existence of something expressed in that idea. It is quite legitimate to press them on the question of how to program the verb "is," this simplest and yet most profound of all verbs. It is quite legitimate to press champions of artificial intelligence about the difference between words as universals and the invariably particular character of things existing. It is legitimate to confront them with a variant of Maritain's observation, that there is infinitely more reality in a cherry in one's mouth than in all the volumes of idealist philosophies.

What Maritain said about idealist philosophies is valid also of all forms of logicism, the kind of philosophy which is blissfully espoused by champions of artificial intelligence. Logicism cannot even account for its being accepted blissfully. Bliss is something related to love, the very factor which moves and agitates human reality much more than intellectual considerations.

About love let me recall here some brief phrases that cast indeed a very long shadow for those who want to reflect. Many of these phrases are exercises in hyperbole. Now a hyperbola, taken for a geometrical figure, is already beyond strict computerization, precisely because its two branches recede into infinity. This realm of the infinite is the realm of pure rational inferences, but not of experience or of strict computational capabilities. In fact the verbal forms of those inferences are such as to justify a remark of Gauss, the prince of mathematicians: "The infinite is only a *façon de parler*."[19] Gauss might just as well have said that the infinite is a metaphor, however valid. Indeed, the impossibility of fully programming a hyperbola is also the impossibility of giving strict contours to metaphors. Metaphors (especially their exaggerated kind, appropriately called hyperboles) are powerful because through them one suggests something which is left intentionally indefinite.

There is simply no room in computers, where everything is strictly definite, and trivially definite in the sense of binary units, for phrases such as the following two Spanish proverbs: "He who loves you will make you weep, but he who hates you will make you laugh," or, "Love is like war, you begin when you like and leave off when you can." The point is that both phrases cease to make sense when taken literally. But how can one instruct a computer not to take a given sequence of letters literally, even if it were possible to program letters as such into computers? Not even the finest microscope would ever see a letter engraved on those iron-silicate crystals that reverse their magnetic orientation with each new instruction but do nothing more. If they love nobody it is only because they know nothing and cannot even flip at will, that is, at their own will.

This is the most tangible reason for the fact that computers are lovable but remain forever unloving. An unwitting proof of this is on hand at least for those who can read between the lines, in *Mind Children*, a book by Hans Moravec, a flamboyant promoter of machine intelligence at Carnegie University in Pittsburgh. The book, which has for its subtitle "The Future of Robot and Human Intelligence," has only one feature to recommend it, namely, that it was published by Harvard University Press. This proves that intellectual perversity is now welcome in publishing houses that claim to cater only to intellectual excellence. What is perverse in that book is not so much its author's making somersaults in logic as his also taking recourse in every paragraph to the trickery of leaving basic terms in studied vagueness. It is child's play to promise robots with intelligence far exceeding that of human geniuses if at the same time one can speak of emotion and consciousness as "nebulous and controversial characteristics."[20]

Of the two, consciousness may appear closer to being a sheer logical construct, yet its conceptual richness far exceeds anything one can find in books about logic. It should be worth recalling that consciousness has for one of its chief characteristics the awareness of the actual moment, or the *now*. Just as freedom is a nightmare in a materialistic outlook, so the reality of the *now* should cause endless nightmares to those infatuated with artificial intelligence, as if such logic exhausted all intelligence. They should try to perceive the farce on reading in Moravec's book about a simulator-equipped robot "which finds itself locked out of its owners's home, its battery charge fading. The robot's simulator will churn through different scenarios furiously for a solution—a combination of actions that will result in recharge."[21]

It would be interesting to ponder the question of whether a solution, unless it is of a purely chemical type, can be equated with the recharging of a battery. Even more interesting should seem the word "furiously" in this context. It is one of the very few contexts where Moravec confronts the question of feelings. The subject index of his book contains only four references to feeling, all of them trivial and very brief. Moravec seems to feel that even those of his readers who believe in artificial intelligence would have second thoughts if reminded too often about the strangely human world of human feelings. One and the chief of those feelings, love, is never mentioned throughout the entire book.

This absence of the word "love" in a book which promises a total recasting of human existence and understanding is not a new phenomenon. Marx debased the word "Liebe" (love) as much as Freud avoided the word "Freude" (joy). Of course, as is becoming increasingly clear, Freud made a total mess of love as well as of those who sought in Freudian psychoanalysis the key to love. Both Freud and Marx were preachers of that shabbiest of all religions, which is the worship of machines, mechanical or psychobiological.

Precisely because they are machines, computers cannot love, lovable as they may be. They are unloving and will remain so forever. Moravec provided a telling proof of this by dedicating his book to his wife as the one who made him whole. If this is what really happened, it merely provided another instance on behalf of the truth of the saying that "love makes one fit for any work." If one then takes one's love for a machine, it is a machine very different from all others. None of them is designed for making one fit to do any work, let alone for doing a *whole* or a *wholesome* work, which is obviously far more than performing this or that very specific and very narrow operation.

That love is superior to any and all human activity has found its supreme form in the perception that "God is love." This is the deepest reason for the fact that such sworn enemies of God and religion as Freud and Marx had no room for love in their ideologies. In both was fulfilled the truth of the saying that "those who despise the word of God, will lose out even on the words of men."[22] Champions of artificial intelligence have already gone much too far in their abuse of human words as they try to turn into an artificial network that greatest human art which is to invent words. They would best atone for their mischief by going whole hog and promising artificial love. Then even morons would awaken from their electronic slumber.

---

[1] Baron d'Holbach, *Système de la nature* (London: 1775), p. 19.

[2] *Pascal's Pensées*, tr. W. F. Trotter, with an introduction by T. S. Eliot (New York: E. P. Dutton, 1958), #82.

[3] Ibid., # 792.

[4] *Oeuvres de Blaise Pascal*, L. Brunschvicg and P. Boutroux (eds.), vol. I, p. 308.

[5] *Pascal's Pensées*, # 340.

[6] "On the True Methods in Philosophy and Theology," in *Leibniz Selections*, ed. Philip P. Wiener (New York: Charles Scribner's and Sons, 1951), p. 69.

[7] Ibid.

[8] See my article, "Socrates or the Baby and the Bathwater," *Faith and Reason* 16 (1990), pp. 63-79.

[9] G. Berkeley, *The Analyst*, in *The Works of George Berkeley*, ed. A. A. Luce and T. E. Jessop (London: T. Nelson, 1948-57), vol. IV, p. 68.

[10] See ch. 4, "Gödel's Shadow," in my *God and the Cosmologists* (1989; new rev. ed., Royal Oak, MI: Real View Books, 1998).

[11] For further criticism of the inconsistencies of Bohr's epistemology, see *God and the Cosmologists*, pp. 138-39.

[12] See chapters 1-6 in my *Science and Creation: From Eternal Cycles to an Oscillating Universe* (1974; 2d rev. ed.; Edinburgh: Scottish Academic Press,   1987).

[13] D. R. Hofstadter, *Gödel, Escher, Bach: an Eternal Golden Braid* (New York: Basic Books, 1979), p. 26.

[14] These words of Von Neumann are printed as a motto on the first page of

volume II of D. E. Knuth, *The Art of Computer Programming* (Reading, MA: Addison-Wesley, 1969).

[15] See my essay, "Determinism and Reality," in *Great Ideas Today 1990* (Chicago: Encyclopedia Britannica, 1990), pp. 267-302.

[16] This gem-like remark is from Whitehead's *The Function of Reason* (Princeton: Princeton University Press, 1929), p. 12.

[17] H. Poincaré, "Sur la valeur objective des théories physiques," in *Revue de métaphysique et de morale* 10 (1902), p. 288.

[18] A. M. Wilson, *Diderot* (New York: Oxford University Press, 1972), p. 577.

[19] *Briefwechsel zwischen C. F. Gauss und H. C. Schumacher,* ed. C. A. F. Peters (Altona: Gustav Esch, 1860-65), vol. 2, p. 269. The letter is dated July 12, 1831.

[20] H. Moravec, *Mind Children: The Future of Robot and Human Intelligence* (Cambridge: Harvard University Press, 1988), p. 44.

[21] Ibid., p. 49.

[22] C. S. Lewis, *That Hideous Strength* (1946; New York: Collier Books, 1962), p. 351. "Qui verbum Dei contempserunt eis auferetur etiam verbum hominis."

4

The Biblical Basis
of Western Science

Science may be a refined form of common sense, but at times all-too refined. Some basic laws of science can, of course, be fully rendered in commonsense terms. The full truth of the three laws of thermodynamics is given by saying that, first, you cannot win; second, you cannot break even; third, you cannot even get out of the game.

Those three laws mean that ultimately all physical activity tends towards an absolute standstill. This is true even if the present expansion of the universe were followed by its contraction. The next cycle of expansion-contraction would be less energetic, and the one after that even more so. Physics, the most exact form of science, tells us, if it tells anything, that all physical processes are part of a one-directional, essentially linear process.

Scientists were not the first to perceive that such is the case. In a more commonsense form it was the Bible that first spelled out this unidirectional process of everything. First, there is creation, then

Address at the meeting of the Philadelphia Society, Philadelphia, April 26, 1997. Reprinted with permission from *Crisis*, October 1997.

cosmic and human history, all tending towards a final judgment and to a final consummation for all in a new heaven and a new earth.

Wherever we find this linear perspective of linearity we find the Bible in the background. This is best appreciated if we take a look at the cosmic view of all great ancient cultures. They are all dominated by the belief that everything will repeat itself to no end, or by the idea of eternal returns. Only on occasion does one hear about this. One hardly ever hears that this belief or idea was responsible for the fact that science suffered a stillbirth, indeed a monumental stillbirth in all ancient cultures.

I coined this phrase, the stillbirths of science, about thirty years ago. The phrase certainly did not catch on in secular academia. The reason is obvious. Nothing irks the secular world so much as a hint, let alone a scholarly demonstration, that supernatural revelation, as registered in the Bible, is germane to science. It is not only germane to science, but it also made possible the only viable birth of science.

That birth took place in the West, in a still Christian West. Modern de-Christianized West still owes its global leadership to that birth which today fuels neocapitalism. The latter needs not only free markets but also merchandise to market, and needs that merchandise in large quantities and in ever new varieties. Only science can deliver them. The rise of that science, so crucial for Western man and for the modern world, has distinctly biblical origins insofar as the Bible is a record of Christian faith.

Whether modern man would be willing to learn in detail about the dependence of science on the Bible is strongly doubtful. But Christians will overlook those details only at grave peril in a great cultural contestation where science plays such a prominent role.

The notion of cosmic linearity, already mentioned, is rooted in the biblical teaching of creation out of nothing. This teaching is not yet present in the classic biblical document about creation, the first chapter of the Book of Genesis, or simply Genesis 1. To read that teaching into that chapter is forgivable in comparison with efforts to see in that chapter something, science, which is certainly not there in any form whatsoever. The sad fact is that nothing has brought so much discredit to the Bible as the chronic effort to take Genesis 1 for a science textbook. It did not start yesterday. But as long as this process goes on, any argument about the biblical basis of Western science will surely boomerang.

Primarily Genesis 1 is not about creation. It is about the importance of the sabbath observance. In Genesis 1 God is set up as a role model who works six days and rests on the seventh. But once God is set up in this role, He is to be assigned the highest conceivable work which is the making of everything.

Genesis 1 states this in three steps, each time using the same metaphor. In English we have the metaphor "lock, stock, and barrel," or the three main parts of a rifle. We often use that metaphor to state literally that we mean everything under consideration. When the Bible states that God made the heaven and the earth, it uses the two main parts of the Hebrew world view, to convey the message that God made everything. The same procedure is repeated in reference to the work done on the second and third days, the special formation of the two main parts, the firmament and the earth. It is with the same thrust that Genesis 1 speaks of the work of the fourth and fifth days, the main decorations of those two main parts. The procedure is to assert that the object of God's work is that totality which is the universe.

The Bible nowhere suggests that the six days can be taken for six geological ages. Nowhere does the Bible suggest that we should read the modern biological notion of species into Genesis 1, where it is stated that God created all the living things according to their kind.

With all that discredit piling up on the Bible through its very first chapter, we should not be surprised that it is well-nigh impossible to sell to secular modern culture a most fundamental biblical message: the total dependence of all on God. In the Bible even the heavens and the stars are on equal footing with muddy earth in that respect. Within the biblical world view it was ultimately possible to assume that the stars or planets and the earth are ruled by the same laws. But it was not possible to do this within the world vision that dominated all ancient cultures. In all of them the heavens were divine.

And the Greeks drew the logic of this with a particular precision, which is the reason why science suffered a stillbirth even among the Greeks of old, those mythical models of modern rationality. Within the Greek ambience it was impossible, in fact it would have been a sacrilege to assume that the motion of the moon and the fall of an apple were governed by the same law. It was, however, possible for

Newton, because he was the beneficiary of the age-old Christian faith.

The faith was Christian in that most fundamental sense, in which the Bible holds Christ to be the only begotten (*monogenes*) Son of God. When faced with that proposition, a well-educated Roman or Greek had his major intellectual shock, apart from shock relating to the moral level. For in Greco-Roman antiquity, the word *monogenes* was an attribute of the Universe itself. This was more than logical in a pantheistic outlook of cosmic emanationism. Therefore such a pagan, ready to convert, had to face up to the following choice: either Jesus or the universe was the only begotten. In other words, Christian faith and pantheism were concretely irreconcilable with one another because of the concreteness of Jesus. This is why only genuine Christian faith, and it alone, can resist the modern juggernaut of nature worship.

The very same belief in Jesus in whom God created everything (and therefore Jesus was God) concretely opposes efforts to make it appear that the universe necessarily is what it is and cannot be anything else. That belief opposes, even when many believers do not realize this, efforts that try to make it appear that the universe necessarily is what it is and cannot be anything else. Such efforts are apt even today to lead science into a blind alley. Two thousand years ago they caused science to suffer a stillbirth among the Greeks of old.

Only one aspect of this intricate subject can be discussed here. It has to do with the Christian, biblical teaching of the creation of the universe in time. God, of course, could have created the world eternally. This is a possibility which neither philosophy nor science can decide on in one way or another. Science could prove the eternity of the universe only if it were possible to perform an experiment that would go on from eternity to eternity. Such an experiment would take a chain of an infinite number of Rip Van Winkles to perform.

The Bible strongly suggests and Christian faith explicitly states that the world was created in time, which means that its past history is finite. How long that history has been, nobody will ever know. We know that physical processes have been going on for at least 15 billion years. But there is no science that can pinpoint that absolutely first moment of existence. For in order to do so, science would have to be able to observe the transition from non-being into being, which is not a physical process.

Science owes to Christian faith in a creation in time the very spark which made Newtonian science possible. That science is based on the three laws of motion. Once those laws were formulated, a science was on hand which from that point on developed in its own terms, with no end to its progress, to its ever new findings, to the ever new merchandise it makes available for the free and at times not so free markets of neocapitalism.

But that irresistible progress needed a spark, the idea of inertial motion, which is the first and most fundamental of Newton's three laws of motion.

The formulation of the first law preceded Newton by more than three hundred years. It first appears in the commentaries which John Buridan gave on Aristotle's book, on cosmology, *On the Heavens,* at the Sorbonne around 1348. By then many other medieval philosophers commented on that book, and radically disagreed with Aristotle's claim that the universe was eternal, that the celestial sphere rotated eternally. The Aristotelian world machine is a perpetual motion machine. As such it blocks the possibility of perceiving an absolute beginning for physical motion. It was, however, this perception that sparked Buridan's insight.

Unlike his many theological predecessors, he did not merely restate the fact of an absolute beginning. He also inquired about the *how* of that beginning. In answer he said almost verbatim: in the beginning when God made the heaven and the earth, he gave a certain quantity of motion to all celestial bodies, which quantity they keep because they move in an area where there is no friction. This is, of course, an uncanny anticipation of Newton's first law, the law of inertial motion. Only after that first law had been formulated was it possible to think about the other two laws.

Secular academia still does its very best to play down the importance of Buridan and of Pierre Duhem, who almost a hundred years ago set forth the evidence about Buridan and medieval science in huge, heroically researched volumes.

Whether a dent will be made on that resistance to the biblical origins of Western science depends, first, on the Bible being read intelligently and, second, on the history of science being studied sedulously. Both are needed if one is to make not so much a spirited but an intellectually respectable case on behalf of the biblical origins of Western science.

In saying intellectually respectable, I also mean biblically genuine. For of all places it is in Paul's Letter to the Romans, this great document on God's grace, that we find the warning: Christian worship must be intellectually respectable. Paul's words, *logike latreia* (Rom 12:1), certainly do not mean logic chopping. But they certainly mean "reasonable," or being "respectful of reason." Why? Because God created men in his own image, an image that certainly includes rationality.

That rationality imposes nothing less than full respect for the ability and rights of reason. This is why already a Saint Augustine laid down the rule that whenever a phrase of the Bible conflicts with what can be known by reason with certainty, it is that phrase that should be reinterpreted accordingly. Otherwise, he said, infidels would raise their laughter sky high and rightly so. The rule of Augustine had already been quietly obeyed in respect to the difference between the Bible's view of the earth as a flat disk and the truth established by Greek science that the earth is spherical.

Unfortunately Augustine himself did not exploit his rule with respect to the firmament, which he blandly located in a vapory layer in the orbit of Saturn. Nor was Augustine's rule heeded when it became imperative, through the work of Copernicus, to attribute two motions to the earth. With an eye on the Bible Martin Luther called Copernicus a fool, and later Rome condemned Galileo, again with an eye on the Bible.

The proper lesson was at long last drawn by the Catholic Church when she left Darwin alone. Darwin is still resisted by many Christians on the ground that God made all plants and animals according to their kind. They resist for the wrong reasons a Darwin who himself failed to realize that the strongest reasons on behalf of evolution were offered by the metaphysical abilities of the human mind which he tried to discredit once and for all. For only that mind can see an interlocking unity across all time and space: from the quarks on to the human body itself, with no gaps in between whatever.

Of course, evolutionary biology is far from having filled all those gaps. Some of them, buried in the past, it may never bridge. But to try to fill those gaps with a recourse to God and to the Bible, would be a most unbiblical thing. First, the history of science has provided countless examples of thus filling gaps of knowledge, each time exposing to ridicule a God whom some ill-advised Christians let perch over this or that gap in scientific knowledge. They took improbabilities for impossibilities, which is an elementary fallacy in reasoning.

One can indeed make an impressive sport of calculating the improbability of this or that physical process. But time and again science performs the "impossible." It should be enough to think of the synthesis of urea by Wöhler in 1828, who in one stroke eliminated the allegedly absolute difference between inorganic and organic matter. After he did that the laughter of some materialists reached high heaven.

Another reason for holding evolution relates to the emphatic affirmations in the Bible that all matter is good. By saying that matter is good, the Bible certainly implies that matter is not evil, but it also says that the edifice raised by God is as good as any other edifice which is good. But an edifice is good only insofar as it is compact, solid, consistent in its working. In other words, such a material edifice fully obeys the rationality of its architect. Why not say all this, and in a superlative sense about the material universe made by God? Is God a second-rate architect, is God a second-rate materials physicist or chemist, or molecular biologist who always has to improve on what he has done already?

Indeed, all the praises accorded by materialists to matter should pale beside the praises which Christians should accord to matter. Herein lies the reason why a Christian should be an all-out material- ist, provided the human mind is excepted. This is why a Christian should be an all-out evolutionist, provided the human mind and the human mind alone is considered as a special creation of God.

Anything short of this would add to the laughter of materialists that reaches to high heaven. I hope that Carl Sagan is now in heaven. So God has the last laugh, that God whose infinite mercy has souls for its object. Even Almighty God cannot be merciful with mere matter. But Carl Sagan has the next-to-the-last laugh. This chief village atheist of our times, or rather the chief atheist performer of the village called evolutionary science, now can laugh fully, knowing that there is no Christian physics, no Christian chemistry, no Christian evolutionary science as long as these are science and not philosophies. But Sagan also laughs at his folly of having promoted the cause of an atheistic science.

This shows that nothing is so dangerous as to latch philosophies to purely quantitative considerations, which are the exclusive business of science. For unless we grant science everything which is its right, we cannot deny anything to science which it cannot rightfully claim.

Nothing which is non-quantitative is the business of science. But everything which is quantitative is its business. Non-quantitative aspects of existence, such as purpose, freedom, design, honesty, cannot be handled by science because they are not quantitative propositions. But every bit of matter is quantitative and therefore the business of science. Does not the Bible say that God "disposed everything according to measure and number and weight" (Wis 11:20)? Please note that the Bible does not say that measure, number and weight, or quantities in short, are everything. But the Bible says that every thing has measure, number, and weight or quantitative properties. Wherever there is matter, quantities are present. This is what gives science its unlimited competence in everything material, whether living or dead. But this is also the reason for the radical limitation of science to what is material insofar as it can be measured.

Herein lies an apparent paradox. It will certainly bother those who do not want to use properly their God-given reason. They do not have to over-exert themselves. It is enough to consider that of the various categories of human conceptualization, there is one which stands utterly apart from the rest. That category is the category of quantities. About all the other categories, various qualities for instance, it is possible to apply the phrase, "more or less." Goodness can be realized in various degrees, more or less. Alertness too. Any food can taste good, more or less. But it is not possible to state about the number five that it is more five or less five.

This profound difference between quantities and any other concepts may not exist for pure spirits and certainly not for God. But it exists for us as long as we are in this mortal body. Chafe as we may, we cannot do anything about the fact that God created the human mind in such a way that for it quantities and everything else remain in two separate conceptual compartments. In other words, what God has separated, no man should try to join, that is, to fuse together. Those busy with integrating theology and science should pause.

There were, of course, some who tried to make it appear that if you pile quantities upon quantities you get qualities and even mind and free will thrown in for good measure. Unluckily for them they tried to write science on that basis but only made a mockery of it and utter fools of themselves. Examples are the Hegelian Right and the

Hegelian Left. They made a horrible mess not only of human life but also of science, including the science of evolution.

It matters not that Darwin's mechanism of evolution is incomplete. It may indeed be grievously faulty. It is always useful to learn about the latest fault lines in Darwinian theory, because its materialist champions love to present it as something scientifically faultless. But this leaves intact Darwin's basic insight. Only those resist it who are inclined to resist either facts or sane philosophy or both. Yet nothing supports evolution so strongly as sane philosophy and especially that biblical precept that everything God made is good and that he arranged everything according to measure, number, and weight. That Darwin failed to see this is largely irrelevant. Without any doubt he proposed his mechanism of evolution as a rebuttal to belief in God, who at that time, and certainly in Darwin's broader ambience, was equated with the God of innumerable special creations.

It was not the first time in intellectual history that God allowed a monumental half-truth so that full truth might be perceived all the more effectively. The half-truth was the combination of an inadequate mechanism of evolution with a magnificent vision of the total coherence of all material beings, together with a much needed radical exclusion of special creation. Darwin's greatest mistake was that he did not take that vision for what it was, a genuinely metaphysical vision.

Metaphysics, and not so much science, is the chief rational basis for stating that the material realm is fully coherent, that is, needs no special interventions from an outside factor, such as God, to keep it running. Science is and will remain profoundly materialistic as long as it is science and not something else. Science can be materialistic only because all matter was created by God. Only a God who is a Creator was capable of giving autonomy to his material creation, without suffering thereby a loss to his omnipotence. Such a God is the God of the Bible.

We shall do the worst disservice to the idea of the biblical origin of Western science as long as we hanker to find in science that something else on the basis of science and in its very terms. For if we take the phrase "according to their kinds" of Genesis 1 in a scientific sense, we have to take everything there also scientifically. What is sauce for the gander is also sauce for the goose. Then we must explain how visible light came before the making of the sun on the fourth day. It is rather ridiculous to claim that the light of the first day was electromagnetic radiation, let alone that it was the 2.7°K cosmic background radiation. Then some explanation has to be found for the firmament and for the astronauts. The Bible deserves much better than to be exposed to endless ridicule by taking it for a science textbook. But the Bible also demands serious intellectual effort if one is to make a case on behalf of its having served as the origin of Western science.

We must make that case partly because the future of Western culture hangs in balance. That culture needs much more than science. We must use both the best means and also the most effective means if we want to obtain a hearing for that much more. A most effective means is nowadays a reference to science. Science, unfortunately, has become one of the three most effective marketing means. The other two are Sports and Sex, writ large. Such are the three S's that rule modern life.

Science, of course, deserves much better and it deserves the best in the way of intellectual efforts. At times it is enough to use common sense. Science may be much more than a refined form of common sense, but in interpreting science correctly some such sense is indispensable. The Bible is an unexcelled source of common sense, and also a chief depository of information about that infinitely much more which is the Kingdom of God. To seek first that Kingdom has been the God-enjoined method of obtaining the rest which, as history shows, includes even science.

For documentation and further discussion of various topics in this address, see my book, *Bible and Science* (Front Royal, VA: Christendom Press, 1996).

5

The Inspiration
and Counter-inspiration
of Astronomical Phenomena

Astronomical phenomena are in a sense mere celestial phenomena, but actually much more. Suffice it to think of the enormous conceptual surplus which is seen in such ordinary celestial phenomena as comets, eclipses, and planetary conjunctions whenever they are seen through the eyes of astronomy and not merely with the naked eye, strengthened as this eye may be by geometry. The surplus in question should seem even more obvious in reference to such extraordinary celestial phenomena as supernovae. Whatever the relative unimportance of their motion through space, their mostly spectroscopic study too rests on the full formulation, since the times of Newton, of the three laws of motion.[1] It is on the application, immediate or remote, of those laws that all physical science, including astronomy, rests.

The text of an invited talk at the International Conference on the Inspiration of Astronomical Phenomena, Castelgandolfo, Italy, June 1994. Reprinted with permission from *Asbury Theological Journal* 51, Nr. 2 (1996), pp. 71-86.

This distinction between mere celestial phenomena and astronomical phenomena bears also on the inspiration which they respectively produce. Let us take the respective reactions to the same kind of phenomena, supernovae, between 1054 and 1987. In June 1054 Chinese stargazers spotted a novel bright spot in the sky which, not surprisingly, they took for a guest star (*kho hsing*), the Chinese name for comets. The fact that it did not infringe on Aldebaran inspired in them the view that the rule of the emperor would be beneficial.[2] Such an inspiration belongs to the class of vain hopes and unnecessary fears triggered by comets and other celestial phenomena listed above. The prospect of removing such fears from the human mind was, in fact, a chief benefit which Halley celebrated in the ode he prefixed to Newton's *Principia*.[3] Inspiring as this prospect could be, it remained for long but a prospect and not a result to be shared broadly.

Considerable improvement in correlating positions, either through naked-eye observations and/or by more refined geometrical methods, did not raise inspiration to a level much higher than the class described above. This is amply revealed in the reaction of Tycho Brahe, the most accurate observer of the sky prior to the advent of telescopes, to the second spotting, in recorded history, of a supernova. Not knowing anything of what those Chinese stargazers had seen in 1054, Tycho Brahe felt that he had made a truly historic first when, on the evening of November 11, 1572, he noticed a very bright star in Cassiopeia. This novelty was in fact the very first item Tycho Brahe mentioned in the long-winded title of his *De nova et nullius aevi memoria prius visa stella . . .* , a book of 104 small quarto pages which he published in short order, excited as he was by what he had seen.[4]

As far as inspiration was concerned, the title of Tycho Brahe's book could seem promising at a superficial look. The new phenomenon, he stated, inspired him to engage in "mathematical contemplation." Of course, this contemplation had nothing to do with the kind of contemplation about which mystics are the best authorities. Tycho Brahe's mental eyes were fixed on the astrological art of predicting the weather from the planets' positions. He felt that once those positions were related to the new star's position the credibility of that art would be greatly strengthened. In other words, Tycho Brahe's inspiration was an increased sense of job-security. Whatever the genuineness of such an inspiration, it certainly had a strong touch of modernity. Tycho Brahe was not, however, so modern as to see in the new star a refutation of the Aristotelo-Ptolemaic doctrine of the incorruptibility of the heavens.

Historians of science would look in vain for traces of some rationalist or iconoclastic modernity in the elegy with which Tycho Brahe introduced his booklet on the new star. The elegy betrays the kind of inspiration which would best be called lucubration. Indeed, this was the very word which Tycho Brahe, at the last moment, decided not to let grace, or rather disgrace, the title page of his hardly inspiring booklet.

Much more modern, and certainly far deeper was the inspiration which Kepler derived from his observations of the nova of 1604. Excited he certainly was. Otherwise he would not have dashed off a book *De stella nova in pede Serpentarii.* But his excitement was that of a deep-seated concern. The new star could easily be taken for a disproof of the starry sphere and for a proof of the presumed truth of the idea that the universe was an indefinite, infinite agglomerate of stars. By 1604 Giordano Bruno had already created some excitement with his strange inspirations about infinite worlds, all forever changing into one another, with no basic difference between stars and planets. Kepler sensed that had Bruno not been burned at the stake in Rome in 1600 (or twenty or so years earlier in Geneva, had he not abjured there his doctrines), he would have seized on the nova of 1604 as a licence for a reckless wandering across infinite spaces. To nip in the bud this kind of use of astronomy and its phenomena, Kepler felt that "astronomy was to be forced to return to its very confines. For certainly nothing good was to be gained by vagabonding through that infinity."[5]

Unlike Kepler, many modern astronomers love that vagabonding.[6] Often they register, with no trace of agonies, their view that the better the universe is known, the more purposeless it appears.[7] However, they fully share Kepler's excitement about the heuristic value of precise measurements. There is something mystical in Kepler's singing the praises of Tycho's measurements as the key to the breakthroughs which later became known as Kepler's three laws and which greatly helped Newton make modern scientific astronomy possible.

Precise measurements were in fact the basic reason for the variety of inspirations which suddenly filled astronomers in late February 1987. On February 24, to be exact, their instruments alerted them to the flare-up, in the southern hemisphere, of a supernova. Apart from the excitement felt over the novelty, the first data inspired them, at least in the sense that they knew they would not soon run out of research problems. The sense of job-security is not something to be taken lightly, even when it is accompanied with the sobering realization that long-standing theories about the origin of supernovae would have to be drastically revised.

Before long this somber mood yielded to a sort of exultation. By June specialists in supernova structure and evolution felt confident that the data dramatically strengthened their theories, worked out over several decades. A reason for this was, as *The New York Times* reported, that already in early March "astronomers and technical experts, usually jealous of their findings, were pooling their observations" as they tried "to solve the supernova's many mysteries."[8] An inspiration certainly commends itself when it helps eliminate selfishness and promotes co-operation. Before long, still another kind of inspiration made itself felt, as leading astronomers took the view that the data pouring in would shed much light on the ultimate fate of the universe.[9]

The history of modern astronomy shows many other cases where these two kinds of elation have been felt over a new astronomical phenomenon, or discovery. One form of elation is felt over the fact that the astronomical phenomenon provides the seal of truth on a theory. The other form is felt over the fact that the astronomical phenomenon opens vistas of further work which may carry the theorist far beyond the range of what has already been worked out. Take, for instance, the discovery by Leverrier, in 1845, of Neptune, a planet postulated by periodic disturbances in the orbit of Uranus. A stunning proof of the truth of Newtonian celestial dynamics, the discovery of a new planet produced so great an elation as to make Auguste Comte decry it as insane. But there was nothing insane in W. C. Tombaugh's heroic work which ultimately led to the discovery of Pluto.[10]

Comte had an axe to grind. Nothing was more dangerous for his positivism than anything really novel in science, especially in astronomy.[11] He did not live to hear the voice of jubilation that greeted Higgins' observation of traces of helium in the spectral lines of the sun. Here too an astronomical phenomenon provided an inspiring capstone on a work already under way, a work initiated by Fraunhofer. Huggins' work also spurred the registering, in vast numbers, of spectral lines. The theoretical coordination of those spectral lines began with Bohr's model of the hydrogen atom. When first told about it, Einstein was inspired to state: "But this is then the greatest of all discoveries."[12]

A capstone on the truth of theories about a very early hot state of the universe was provided by the discovery of the 2.7°K cosmic background radiation in 1965. The ensuing excitement went hand in hand with the inspiration to do further and extensive study, theoretical as well as experimental, on that radiation. But the inspiration had other aspects as well. Such a hot early state could not be reconciled with the steady-state theory. While this and other consequences of that radiation exhilarated the proponents of what by then had been known as the Big Bang, it inspired the grim resolve of the champions of the steady-state theory to keep working out alternatives to an apparent cosmic beginning.

Champions of the steady state theory disclosed only now and then that their opposition to the Big Bang was motivated by a counter-theological inspiration. The rank materialism of most champions of the steady-state theory dictated that the Universe was the ultimate entity, and as such it had to be without a beginning. Unfortunately, only on rare occasions was that materialistic inspiration exposed by prominent astronomers. One of the few was Arno Penzias, co-discoverer of that radiation.[13] He was, however, hardly right in buttressing the opposite kind of inspiration with his claim about Genesis 1. In the phrase of Genesis 1, "Let there be light!" Penzias saw an anticipation of the 2.7°K cosmic background radiation.[14] This was a most unfortunate echo of that blind inspiration which animates those who nowadays refer to themselves as creationists, that is, those who take Genesis 1 for a science textbook.

Counter-theological, or strictly materialistic, inspiration is not absent either in the effort to find the so-called missing matter. In itself the effort is purely scientific. Clearly, not enough matter is known to exist if the rotational dynamics of galaxies obeys Kepler's third law. But one wonders whether non-scientific inspiration is not strongly at work in sustaining the search for the missing matter.[15] A successful outcome of that search would be taken by not a few as a proof of an eternal cosmic dynamics. Had not such motivations been at work, less despondency would have greeted the news that the very first experiments completed with the Keck Telescope in Hawaii, the largest telescope in the world, yielded evidence about a surprisingly large abundance of deuterium in the distant, and therefore early, universe. This was inspiring news for advocates of the Big Bang, but very bad news for those searching for the missing matter, let alone for rearguard advocates of the steady-state theory.

For both camps there remained, of course, the excitement or inspiration, about the prospect of vastly improved observational possibilities. This is what John N. Bahcall seemed to emphasize in saying that the result in question had astronomers "dancing in the dark corridors of their observatories." The new telescope was so great a success, he continued, that "some of the questions that astronomers have sought to answer for decades may be solved in a night's observations with these new eyes."[16]

It is not, however, easy to keep from view the religiously colored inspiration in reference to even the latest astronomical phenomena. A case in point is George Smoot's announcement of slight variations in the 2.7°K cosmic background radiation. The news produced an outburst of reactions, many of them inspirational in a religious sense. Smoot himself first took the view that, to quote his very words, "if you are religious, it's like looking at God." A week or so later, being reminded of this, Smoot tried to balance that religious inspiration with a distinctly secularist one: "What matters is the science; I want to leave the religious implications to theologians and to each person, and let them see how the findings fit into their idea of the universe."[17]

Underlying this balancing act is the fact that one and the same astronomical phenomenon can generate inspired states of mind which, differ as they may from one another, subjectively can be designated by the same word, inspiration. This can happen even when the same religious sentiments are intensely shared. While John Donne was downcast by the apparent vanishing of all coherence because of the rise of heliocentrism and atomism, both were taken by Pierre Gassendi, also in holy orders, for harbingers of good news. More wisely, Pascal, greater than those two as a philosopher, as a scientist, and as a Christian, argued that science is absolutely impotent to deliver even a drop of that supreme inspiration which is genuine selfless love.[18]

Of course, when religious sentiments, let alone the same religious sentiments, are not shared, it is almost inevitable that the same astronomical phenomenon will produce widely different inspirations. One such difference became a legendary page in the history of astronomy. To the question of Napoleon, who had found no reference to God in Laplace's *Système du monde*, Laplace answered that he did not need that hypothesis. While within science proper this was a most defensible position, Laplace conveyed something of the practical atheism which animated him in those years. Indeed, countless writers and speakers took his words for a proof that atheism or agnosticism is the inspiration appropriate to the science of astronomy. It is rarely mentioned that when Laplace uttered those memorable words, Herschel was present and politely disagreed.[19]

There was no such confrontation between the Abbé Lemaître and Robert Millikan as they served on the panel which the British Association sponsored in 1931 on the latest in cosmology, the expansion of the universe and, by implication, its origin. This is not to suggest that the confrontation was not a distinct possibility and all the more so as both Lemaître and Millikan distinguished themselves with work on cosmic rays and both saw the question of the origin of cosmic rays as relevant to the question of the origin of the universe. Far from agreeing with the suggestion that perhaps "an infallible oracle" might provide the answer, Lemaître preferred the oracle's silence so that "a subsequent generation would not be deprived of the pleasure of searching for and of finding the solution."[20] Millikan, however, suggested that if theories proved that annihilation processes went on in interstellar spaces and not only within the stars themselves, this would "obviously influence strongly not only present theories but also all future theories of the origin and destiny of the universe."[21] What Millikan expected was nothing less than a scientific proof of the view, hardly verifiable scientifically, that matter is eternal. Unlike Lemaître, who kept his philosophico-religious inspiration apart from doing science, Millikan readily grafted on science a counter-religious inspiration.

While Millikan, and like-minded scientists in the West had the freedom not to do so, scientists in the Soviet Union were forced to mix their scientific inspiration with a materialistic counter-inspiration imposed on them. I personally witnessed of one such case, the last-minute appearance of V. A. Ambartsumian as member of the cosmology panel at the 17th World Congress of Philosophy in Düsseldorf, in 1978. There, in order to reward the Party for the opportunity to go abroad, he suddenly departed from his topic dealing with stellar evolution. He did so in order to declare in a phrase or two that no scientific conclusion had a better empirical foundation than the doctrine of dialectical materialism about the eternity of matter.[22]

But even when non-scientific sources of inspiration are kept out of focus, the inspiration sparked or sustained by work on much the same astronomical phenomena can reveal differences worth noting and all the more so because they clearly point beyond what is strictly scientific. Edwin Hubble concluded his classic *The Realm of the Nebulae* in words where a grim resolve to continue the exploration of space is coupled with a scorn for theoretical reflections. What made Hubble scorn theories was not, however, his love for experimental work, but his infatuation with empiricism. At the dim boundary, or the utmost limit of our telescopes, Hubble wrote, "we measure shadows, and we search among ghostly errors of measurements for landmarks that are scarcely more substantial. The search will continue. Not until the empirical resources are exhausted, need we pass on to the dreamy realms of speculation."[23] A philosophically very different end-note was struck by Richard C. Tolman in his equally classic *Relativity, Thermodynamics and Cosmology*: "It is appropriate to approach the problems of cosmology with feelings of respect for their importance, of awe for their vastness, and of exultation for the temerity of the human mind in attempting to solve them. They must be treated, however, by the detailed, critical, and dispassionate methods of the scientist."[24]

The contrast between these two grand conclusions should seem all the greater as both were first published about the same time, the mid-1930s. The second came from a leader in a highly theoretical relativistic cosmology who clearly relished the inspirational power of theorizing. The first came from one so disdainful of theories as to fail to acknowledge that all empirical observations are theory-laden and, indeed, to so great an extent as to beckon to domains open only to eyes inspired by much more than mere science. To treat with empiricist contempt such domains is no less mistaken even from the purely astronomical viewpoint than to wade into their vast reaches with the presumption that scientific skill is enough to do philosophy and theology even moderately well.

One need, however, be on guard against believing that the consideration of the history of astronomy may readily impose a fair measure of sobriety on students of the realm of the stars, nebulae, and, indeed, of the astronomical universe as such. And what if sobriety begins to parade in the garments of that subtle dizziness which is known as solipsism? Two highly regarded surveys of the history of twentieth-century cosmological theories fully illustrate this fearful outcome. For whatever the markedly pragmatico-idealistic philosophies of their respective authors, the theories surveyed by them provide ample material for supporting their doubts about the reality of the astronomical universe. They, however, have failed to see that by taking the latent or unabashed solipsism of many a cosmologist and astronomer for science they not only do rank injustice to the cosmos or universe but also cast doubt on the merit of the very titles of their books. For if one cannot have rational assurance of the reality of the totality of things which is the universe, there is clearly no such a thing as "the measure of the universe."[25] On the same supposition cosmic reality can but degenerate into "an invented universe" impossible to invent for the purposes of any science that cannot take its instruments for mere inventions.[26]

The author of *The Invented Universe* found that all modern cosmology tends to substantiate W. De Sitter's prediction that the universe is but a hypothesis which "may at some future stage of the development of science have to be given up, or modified, or at least differently interpreted."[27] This dispiriting prophecy can be seen to come true in that incoherent statistical ensemble into which the coherent totality of things, or Universe, is turned in quantum cosmological theories. Their proponents are signally oblivious to the fact that conviction about the rational coherence of all things, however distant from one another, has from the start been the great inspiration which propelled science, including astronomy. The inspirational lifeblood of astronomy depends indeed on giving a firmly affirmative answer to the question, *Is There a Universe?,*[28] having taken that universe for the strict totality of consistently interacting material entities.

These details, old and new, from the history of astronomy put one face to face with a wide variety of meanings which the word inspiration may carry. Therefore it may be worthwhile to take a close look at the word itself. Otherwise this conference[29] too may suffer the fate typical of almost all of them. All symposia, so goes a slightly sarcastic remark, begin in confusion and end in confusion, though on a much higher level. Those who have already sat through a dozen or so symposia will hardly disagree with this far-from-flattering generalization. If one looks for the reason, one may find it in the failure of the organizers to call for a clear definition of basic terms. Or one may find it in the speakers' unwillingness to come clean. In modern academia, haziness, couched in convoluted language, has come to be taken for profundity.

Will that haziness be dissipated by consulting the 18-volume *Oxford English Dictionary*? On a first look, the effort may seem promising. The word inspiration, together with its verb form, to inspire, takes up five columns, or ten times the mere half a column which is the average space allotted to the 400,000 words listed in the 8,000 quarto pages of that truly magnum opus.

Should one therefore expect that the hundred or so uses of the word inspiration listed there stand for a great variety of meanings? Far from it. All those meanings fall into three distinct groups, of which one, the physical act of breathing, or to breathe air into something, may be conveniently ignored for our purposes, unless boredom or the summer heat calls for in-spiration, that is, artificial respiration, or mouth-to-mouth resuscitation. Another meaning of the word inspiration is related to God who supernaturally inspires some thoughts or courses of action. This essentially theological meaning may also be ignored, at least for the moment. Of immediate interest is the third, or essentially figurative meaning. All the varieties of that meaning, filling most of those five columns, hinge on the last word of its definition: Inspiration is "a breathing in or infusion of some idea, purpose etc. into the mind; the suggestion, awakening, or creation of some feeling or impulse, esp. of an exalted kind."[30]

Inspiration is then connected, as was already surmised in the beginning of this paper, with a state of exaltation. Unfortunately, about that state the same vast dictionary does not offer the kind of enlightenment which is clarity. We are told that one is having an exalted thought when, figuratively speaking, one takes some higher ground or perspective. Herein lies hidden a sort of tautology, something even worse than a mere paradox. Taking a higher ground means exaltation which in turn is the principal ingredient of inspiration. Conversely, once one is inspired, one is exalted and therefore on a higher ground. One may indeed ask: When using the words excitement and inspiration are we not running in a circle? Is not the luxury of having two or three different words at our disposal a mere cover-up for intellectual poverty when those three terms— excitement, inspiration, and higher ground—define one another?

This hardly enviable situation is made worse by the fact that in reference to the state of excitement the dictionary makes no mention of the fact that in such a state one is usually animated with a strong sense of purpose, or at least by an illusion of it. In view of this connection, one is entitled to say that an absence of a sense of purpose, a sense of being lost, would be on hand whenever excitement would yield to its opposite, namely, dejectedness or despondency. Therefore one could just as well coin a new word, counter-inspiration, a word not listed in that huge dictionary, although it lists many composite words that begin with "counter-" and, assuming their meaning to be obvious, does not give their definition. Counter-inspiration would then mean to feel not only very low or dejected, but also to feel deprived of a constructive or abiding sense of purpose, or even more picturesquely, to feel oneself to be mere flotsam and jetsam on unfathomable cosmic waters. More of this shortly.

But first the so-called higher ground. It is a treacherous ground when claimed by science and scientists, even when they merely talk of being inspired. Charles Darwin fully recognized this when in 1845 he set himself the rule, "Never use the words higher or lower!" Darwin himself disregarded this rule more often than not.[31] At any rate, the rule meant that even though a monkey should seem to occupy a ground much higher on the evolutionary scale than a mouse, let alone a mollusc, no biologist should call one higher and the other lower for a simple reason: Such a grading is a kind of value judgment which has no place in empirical science.

Even more applicable should seem the same rule in the field of exact physical science, of which astronomy is a principal branch. Unlike biology, or life science, which deals with flesh and blood organisms, of which one is patently more complex and powerful than another, astronomy, like physics, is a systematic leveler. It deals only with lifeless entities, and is interested only in the quantitative properties of their motions. There is nothing higher or lower there, only bigger and smaller, longer and shorter, farther or nearer, but never anything that in purely astronomical terms could be seen evocative of "nearer, my God, to thee," or even of nearer to you, O man.

Such is at least the case as long as we define physical or astronomical science as was done above. On more than one occasion I have felt it appropriate to define physical science as the quantitative study of the quantitative aspects of things in motion. The reason for this was my resolve to save the sciences and the humanities from mutual encroachments and, if I may add, leave whatever inspiration they may offer, in compartments that are at least methodologically separate.

Since I doubt that Leon Lederman share that resolve of mine, I was all the more pleased to find in his book, *God Particle*, a very similar definition of physical science: "Physics is a study of matter and motion. The movement of projectiles, the motion of atoms, the whirl of planets and comets must all be described quantitatively. Galileo's mathematics, confirmed by experiment, provided the starting point."[32]

Lederman's words are a combination of plain truth, of a rank half-truth, and of some basic assumptions that cannot be justified by physics, but without which physics (or astronomy) hangs in mid-air. The plain truth is that unless physics gives a quantitative account of what it deals with, it is not yet physics. The half-truth relates to Galileo's mathematics. It was not mathematics but, as Duhem showed already in 1913, a long medieval tradition that gave Galileo the idea of that accelerated motion which is the only kind of motion, be it the free fall of bodies, that obtains in the real world.[33] Moreover, it was neither mathematics nor geometry that assured Galileo in the first place that matter and motion invariably lend themselves to quantitative considerations. One could, of course, delight, as Galileo did, in the marvelous coherence of mathematics and become greatly excited on that score.

But what assumption justified for him the application of quantities to the physically real? Certainly not the quantities themselves, for this would be a begging of the question itself, a *petitio principii*. The justification can be made only on the basis of assuming that the human mind can know matter and motion, before saying anything quantitative about them. The justification would also imply the tacit acknowledgment that the human mind can validly talk about the totality of quantitatively coherent physical matter which is the universe. At any rate, Galileo found his justification with an eye on the Creator: nothing showed so much the excellence of the Creator than that created human mind with its ability to know quantities as "objectively" as God himself did.[34]

Such an inspiration was fraught with great perils. Galileo indeed claimed that quantities alone counted, and all secondary qualities (taste, colors etc.) had to be considered as purely subjective.[35] The uninspiring cultural results are too well known to be detailed here. In sum, if Galileo's claim is correct one may just as well write off all humanities and take the plague of scientism for a sign of health. Since to the spread of that plague not a few great scientists gave, at least in recent times, unwitting help, the most effective antidote against it may best be sought in statements made by other eminent scientists.

The most impressive of those statements may be the one by Eddington, partly because of its succinct character. The line between the sciences and the humanities does not run, Eddington wrote, "between the concrete and the transcendental but between the metrical and the non-metrical."[36] This remark, carried to the four corners of the scientific and academic world, did not inspire a new climate of thought, although it should have. Yet only by keeping in mind that boundary is it possible to distinguish two kinds of ingredients in the inspiration felt by an astronomer about astronomical phenomena. Some ingredients are scientific, such as the mathematical simplicity of the explanation. Some other ingredients, which are often far more decisive, have nothing to do with the science of astronomy but almost everything to do with the ideology or religion, or perhaps the plain counter-religion, of the astronomer.

For unless that distinction is made, there remains no remedy for a cultural disaster in the making. It is the flooding of the societal scene with the kind of inspiration dervied from astronomical phenomena which is a rank counter-inspiration, in the sense defined above. A notorious example is a passage by a prominent humanist who clearly had no confidence in his métier, which deals with the non-metric in human reflections. I mean Carl L. Becker, a leading American historian of the Enlightenment. To make matters more revealing, most readers of his *The Heavenly City of 18th-Century Philosophers* have been more shocked by a factual truth than by a thorough misinterpretation of some very scientific facts. The factual truth was that the gurus of the Enlightenment were led not by reason but by a dream about heaven on earth. The misinterpretation of the facts is best given in Becker's own words, spread by now through more than thirty printings in twice as many years:

Edit and interpret the conclusions of modern science as tenderly as we like, it is still quite impossible for us to regard man as the child of God for whom the earth was created as a temporary habitation. Rather we must regard him as little more than a chance deposit on the surface of the world, carelessly thrown up between two ice ages by the same forces that rust iron and ripen corn, a sentient organism endowed by some happy or unhappy accident with intelligence indeed, but with an intelligence that is conditioned by the very forces which it seeks to understand and to control. The ultimate cause of this cosmic process of which man is a part, whether God or electricity or a "stress in the ether," we know not. Whatever it may be, if indeed it be anything more than a necessary postulate of thought, it appears in its effects as neither benevolent nor malevolent, as neither kind or unkind but merely as indifferent to us. What is man that the electron should be mindful of him! Man is but a foundling in the cosmos, abandoned by the forces that created him. Unparented, unassisted and undirected by omniscient or benevolent authority, he must fend for himself, and with the aid of his limited intelligence find his way about in an indifferent universe.[37]

The entire passage is a valuational misinterpretation of facts, well established by science, and a presentation of some assumptions as if they were integral parts of science. One needs merely replace the ether with zero-point oscillations in the vacuum, the electron with Higgs bosons, the ice ages with periodic extinctions of life on earth, and Becker's passage would be wholly up-to-date as well as wholly misleading with its counter-inspirational fallacies. No astronomer or cosmologist of note is known to have protested the above passage, which countless undergraduates have had to swallow for the past sixty years. If prominent humanist admirers of Becker have found fault with his book, it has not been for his falling victim to an ideology which claimed him as he penned that passage.[38]

The ideology still works, although Pascal had already unmasked it three and a half centuries ago. He did so as he described the haplessness of the libertine, that is, of the agnostic or sceptic who, in looking out into the vast depths of the cosmos, was terrified by mere distances.[39] Pascal could have also remarked that already the Aristotelian universe was vast enough to unsettle those who sought comfort in short distances. Publications of prominent recent interpreters of astronomy can at most put a brave face on the terror they conjure up as they try to discredit common sense perception with intimations of the unimaginable magnitude of millions and billions of light years. They merely trigger misguided bewilderment.

The result is a feeling of utter dejectedness about being "lost in the cosmos," to recall the title of a much ignored book of Walker Percy's. As a sane novelist, unwilling to play to the galleries, Percy put his finger on that sensitive spot which can never float into the focus of any telescope or microscope. That spot was fully alive in a Mount Wilson astronomer's wife who divorced him on the ground of "angelism-bestialism." The source of this strangely hybrid trait derived from a travesty of inspiration which the astronomical phenomena known as quasars had sparked. The astronomer in question, Percy reported, was "so absorbed in his work, the search for the quasar with the greatest red-shift, that when he came home to his pleasant subdivision house, he seemed to take his pleasure like a god descending from Olympus into the world of mortals, ate heartily, had frequent intercourse with his wife, watched TV, read Mickey Spillane, and said not a word to wife or children."[40]

Clearly, in this case (and many others could be quoted) nothing was gained in the way of genuine inspiration by stretching the limits of the known universe from a few light years to billions of times that amount. But the root of the loss of true inspiration lies not in the observatories. It lies with Christians "in whose eyes the traditional Christian content and promise had become 'absurd'." Such is the diagnosis of Hannah Arendt, an agnostic Jew. She also notes the laughable character of the excuse that either the atheism of the eighteenth century or the materialism of the nineteenth offered serious arguments against that content and promise. Those arguments, she notes, were "frequently vulgar and, for the most part, easily refutable by traditional theology."[41]

Whether Pascal was just as antirational as was Kierkegaard, both of whom Arendt blames equally for the introduction of the Cartesian *dubito* into religious belief, is a minor issue. The principal issue is that the Cartesian *dubito* set up mathematical logic as the only reputable form of cogitation. Precisely because of this, there followed a growing distrust in man of his direct registration of external reality, be it a physical or an historical event. Among the results was a disregard for the factual historical origin or birth of science. In place of facts, myths came to be cultivated by historians of science. When Bergson wrote that science, the daughter of astronomy, "has come down from heaven to earth along the inclined plane of Galileo,"[42] he failed to realize that he had given an inimitably concise rendering of one such myth.

The origin of science had indeed much to do with heaven, though with a distinctly Christian one, anchored in unique facts of salvation history. This is why the question of the origin of science has been a very upsetting topic for many a historian of science.[43] The counter-inspiration which exudes from their accounts of scientific progress has much to do with that unease of theirs.[44] But the disregard of the true origin of science meant also a disregard for the true source of inspiration that liberated science from its repeated stillbirths and provided its only viable birth.[45] Of course, once the basic laws of physical science were in place, it could further develop in terms of its purely scientific attainments, with no consideration for the inspirational spark of its origin.

The spark was belief in creation out of nothing and in time. This belief, because of the status assigned to Christ, worked within Christianity as a unique antidote against the pantheism which caused the stillbirths of science in all great ancient cultures and nipped in the bud the prospect of science even in the medieval Muslim context.[46] This is not to suggest that today one needs to be a Christian to do physics or astronomy or cosmology worth a Nobel Prize or two. But if the same physicist wants inspiration which is much better than Cartesian "angelism" or Darwinist "bestialism," or both fused into one, he or she will have to look in the direction specified by Arendt.

What happened to the attitude toward external reality should be of no less interest as far as the inspiration and counter-inspiration of astronomy is concerned. The *cogito ergo sum*, which was Descartes' resolution of the *dubito*, reached its ultimate unfolding in the principle, "I think, therefore the Universe is,"[47] a half-hearted spoof of the anthropic cosmological principle. The principle cannot do harm to the universe, but it is already ruining the minds of some professional stargazers and cosmologists. The same should seem to be true of quantum cosmology if it suggests that expertise with it enables one to create entire universes literally out of nothing.[48] Such an inspiration is the kind of hubris that opens wide the abyss of sheer counter-inspiration.

If history is a proof, latter-day astronomers and cosmologists will have no genuine inspiration, either for themselves or for the millions who gobble up their non-astronomical words of wisdom, unless they keep an eye on a very specific God. I mean the God whose very first recorded act in Genesis 1 was to let his breath, his inspiration, float over what was to become a universe. This remark may sound like plain sermonizing. If it does, I refuse to apologize. To support my refusal I could recall a Copernicus, a Galileo (yes, a Galileo!), a Newton for all of whom belief in the Creator of the astronomical universe was a signal source of inspiration in their search for a better scientific account of the starry sky. Why, one may ask with Galileo, was Copernicus so inspired as to be willing to commit a rape of his very eyes?[49] I hope that such and many similar details of the history of astronomy are not entirely unknown. At least they can be learned by anyone ready to consult well translated classics of its history.

Here, to support my refusal to apologize I would put the emphasis on a book which is the furthest possible cry from Christian, let alone from Roman Catholic, sermonizing. I mean Sigmund Freud's *Civilization and its Discontents*. To be sure, Freud still refrained from describing the Catholic Church as "the implacable enemy of all freedom of thought" which "has resolutely opposed any idea of this world being governed by advance towards the recognition of truth!"[50] While Freud could not be blamed for having been unaware of the Christian sources of belief in progress, he had no excuse for ignoring Bury's memorable unmasking of the secularist idea of progress as a mere begging of the question.[51] At any rate, in *Civilization and its Discontents* Freud stated that "only religion can answer the purpose of life."[52] Not that he viewed the answer of any religion as satisfactory. Far from it. But, implicitly at least, he ruled out science, even his own science (or rather pseudoscience) of psychoanalysis, as a source of an answer about purpose. The most science could do was to palliate the discontent for some, though hardly for all.

If around 1930 Freud could be struck by a high-level of discontent in our increasingly scientific civilization, one wonders whether he would not be literally dumbstruck today. As an antidote to that grave discontent not a few astronomers, relatively greater in number than say thirty years ago, offer science. Carl Sagan is a prime example. He and others have hoped—perhaps against hope—that with more science there will be less religion. They have all shared something of the delusion, memorably voiced by Herbert Spencer in 1850, that once science-oriented education is universal, equally universal will be the disappearance of crime.[53] Actually, crime is becoming universal, owing in no small part to the misuse of tools provided by science and technology.

Today, we have more science than ever and more scientific education than ever, but also a crime rate which is skyrocketing. Partly because of this we have much more religion as well. The reason for this is the unquenchable hunger of mankind for a sense of purpose that can carry one through crimes and tragedies, as well as abide even beyond that disaster which is the grave. No talk, however exquisite in its rhetoric, about cosmic brotherhood or a biocentric universe proves indeed to be of any personal comfort when, say, a promising young man puts his shotgun into his mouth and blows his brain to pieces.[54]

Science has failed, miserably failed, to still that hunger for purpose. Not that it has ever been its task to do so. The task of science has indeed been greatly compromised by ever recurring efforts of scientists, especially during the last half century, to force science to give what it cannot deliver.[55] But if scientists fail to gain a sense of abiding purpose from a source other than science, their scientific inspiration may not rise higher than the level of feeling some excitement. From there it is but a short step to what I have described as the lowlands of counter-inspiration. Would that ever fewer would present it as the higher ground, let alone as genuine inspiration. As to those who are seized by such an inspiration, may they never lose heart to keep breathing it into many far and wide.

[1] This emphasis on the *full* formulation should be a reminder of its long prehistory. Newton must have known that he owed the second law (action equals reaction) to Descartes who, in turn, could not be unaware of the late medieval origins (unearthed and documented by Pierre Duhem early in this century) of the first law. Thus Copernicus had relied on the notion of inertial motion given in terms of an initial impetus as an idea too familiar to his readers to demand justification or explanation.

[2] Or in the words of the Emperor's Chief Calendrical Computer, "The guest star does not infringe upon Aldebaran; this shows that a Plentiful One is Lord, and that the country has a Great Worthy." Quoted in J. Needham, *Science and Civilization in China.* vol. III. *Mathematics and the Sciences of the Heavens and the Earth* (Cambridge: University Press, 1959), p. 427.

[3] "Now we know / The sharply veering ways of comets, once / A source of dread, no longer do we quail / Beneath appearances of bearded stars." *Sir Isaac Newton's Mathematical Principles of Natural Philosophy and His System of the World,* ed. F. Cajori (Berkeley: University of California Press, 1962), p. xiv.

[4] For a full title as well as a detailed discussion of the book's content, see J. L. E. Dreyer, *Tycho Brahe: A Picture of Scientific Life and Work in the Sixteenth Century* (1890; New York: Dover, 1963), pp. 38-69.

[5] Published in 1606. For details, see my *The Paradox of Olbers' Paradox* (New York: Herder & Herder, 1969), pp. 30-34.

[6] And love it to an astonishing extent. A century ago Lord Kelvin declared finitude to be incomprehensible, while at the same writing off all infinity beyond the confines of our Milky Way as of no physical consequence. (For details, see my *Paradox of Olbers' Paradox,* pp. 168-70). Inattention to basic mathematical and logical problems inherent in the notion of the physically infinite mars F. J. Dyson's Gifford Lectures, *Infinite in All Directions* (New York: Harper & Row, 1988).

[7] I have in mind Steven Weinberg's concluding words in his *The First Three Minutes*, for which he offered a lame apology in his *Dreams of a Final Theory* (New York: Pantheon Books, 1992).

[8] See *The New York Times*, June 16, 1987, Section III, p. 1. and March 8, 1987, sec. I, p. 1.

[9] The scientists quoted to that effect were John N. Bahcall and Sheldon L. Glashow. See *The New York Times* April 3, 1987, p. 16.

[10] Heroic, indeed, as it implied the scanning, with a mere magnifying glass, of hundreds of thousands of photographic plates over a period of thirty years.

[11] The reason for this was Comte's ambition to formulate a scientifically definitive form of sociology. Obviously, then, the prospect of new major breakthroughs in physical science had to appear most upsetting for Comte. For details, see my *The Relevance of Physics* (Chicago: University of Chicago Press, 1966; new ed. Edinburgh: Scottish Academic Press, 1992), pp. 468-77).

[12] As reported by G. Hevesy in a letter of October 14, 1913, to Rutherford. See A. S. Eve, *Rutherford: Being the Life and Letters of the Rt. Hon. Lord Rutherford, O. M.* (Cambridge: Cambridge University Press 1939), p. 226.

[13] In an interview with J. Dietch, "Tackling a Question: Why a Universe?" *The New York Times*, December 1, 1991, p. NJ 3.

[14] See *The New York Times,* March 12, 1978, p. 1.

[15] For details, see my *God and the Cosmologists* (Edinburgh: Scottish Academic Press, 1989), pp. 70-75.

[16] *The New York Times,* April 15, 1994, p. 22

[17] See *The New York Times*, April 24, 1992, p. 1 and 16 and May 5, p. C9.

[18] "All bodies together, and all minds together, and all their products, are not worth the least prompting of charity. This is of an infinitely more exalted order." See *Pascal. The Pensées*, tr. J. M. Cohen (Penguin Classics, 1961), p. 284 (#829).

[19] See my *Planets and Planetarians: A History of Theories of the Origin of Planetary Systems* (Edinburgh: Scottish Academic Press, 1978), pp. 127-28.

[20] *British Association Report for the Advancement of Science. Report of the Centenary Meeting. London, 1931 September 23-30* (London: Office of the British Association, 1932), p. 608.

[21] Ibid., p. 597.

[22] See my *God and the Cosmologists,* p. 61.

[23] E. Hubble, *The Realm of the Nebulae* (New Haven, CT.: Yale University Press, 1936), p. 202.

[24] R. C. Tolman, *Relativity, Thermodynamics and Cosmology* (Oxford: Clarendon Press, 1934), p. 488.

[25] J. D. North, *The Measure of the Universe: A History of Modern Cosmology* (Oxford: Clarendon Press, 1965).

[26] P. Kerszberg, *The Invented Universe* (Oxford: Clarendon Press, 1989).

[27] W. De Sitter, *Kosmos* (Cambridge, MA: Harvard University Press, 1932), pp. 133-34.

[28] The title of my Forwood Lectures, given at the University of Liverpool (Liverpool University Press, 1993). There I offer a proof of the reality of the universe, a proof which, steeped as it is in considerations about quantities, is strictly philosophical.

[29] Held on the theme, "The Inspiration of Astronomical Phenomena," in Castelgandolfo (Italy), June 25-28, 1994.

[30] See p. 1036, col. 3, in volume 5.

[31] Darwin wrote those words on a slip of paper which he kept in his copy of Chambers' *Vestiges of the Natural History of Creation* (1844), a book which presented evolution as a God-directed process toward ever higher forms of life.

[32] L. Lederman (with Dick Teresi), *The God Particle: If the Universe is the Answer, What is the Question?* (Boston: Houghton Mifflin Company, 1993), p. 71.

[33] Or to quote Duhem: "By the middle of the 16th century, Parisian scholastics had considered as banal these truths: The free fall of a body is a uniformly accelerated motion. The vertical ascent of a projectile is a uniformly retarded motion. In a uniformly changed motion, the path traversed is of the same length as its length would be in a uniform motion of the same duration, whose velocity would be the mean between the two extreme velocities of the uniformly changed motion. . . . In favor of these laws Galileo could provide new arguments, drawn either from reason, or from experience; but, to say the least, he did not have to discover them." *Études sur Léonard de Vinci. Troisième Série. Les Précurseurs parisiens de Galilée* (Paris: Hermann, 1913), pp. 561-62.

[34] See Galileo Galilei, *Dialogue Concerning the Two Chief World Systems— Ptolemaic & Copernican*, tr. S. Drake (Berkeley: University of California Press, 1962), p. 103.

[35] "I think that tastes, odors, colors, and so on are no more than mere names so far as the object in which we place them is concerned, and that they reside only in the consciousness." *The Assayer* in *Discoveries and Opinions of Galileo*, tr. S. Drake (Garden City, NY: Doubleday, 1957), p. 274.

[36] A. S. Eddington, *The Nature of the Physical World* (1928; Ann Arbor: The University of Michigan Press, 1958), p. 275.

[37] First published in 1932 by Yale University Press. My reference is to the 27th printing in 1965. See pp. 14-15.

[38] The blindness of those humanists is unwittingly documented in the symposium held on the 25th anniversary of the publication of Becker's classic at Colgate University on October 13, 1956, and published under the title, *Carl Becker's Heavenly City Revisited* (Ithaca, NY: Cornell University Press, 1958). Only one of them, R. R. Palmer of Princeton University, called attention to the counter-theological basis of Becker's despairing outlook, and even he did not

want to appear to have endorsed thereby a genuinely supernatural Christian perspective.

[39] *Pascal. The Pensées,* p. 57 (#91).

[40] Walker Percy, *Lost in the Cosmos: The Last Self-Help Book* (New York: Washington Square Press, 1983), p. 116.

[41] H. Arendt, *The Human Condition* (1958; Garden City, NY: Doubleday, 1959), p. 291.

[42] H. Bergson, *Creative Evolution,* tr. A. Mitchell (New York: Modern Library, 1911), p. 364.

[43] As discussed in my Fremantle Lectures, Oxford University, *The Origin of Science and the Science of its Origin* (Edinburgh: Scottish Academic Press, 1968).

[44] See in particular ch. XIV, "Paradigms or Paradigm," in my Gifford Lectures, *The Road of Science and the Ways to God* (Chicago: University of Chicago Press, 1978).

[45] See on this my book, *Science and Creation: From Eternal Cycles to an Oscillating Universe* (1974; new ed.; Edinburgh: Scottish Academic Press, 1986).

[46] See my Wethersfield Institute Lectures, *The Savior of Science* (Washington, DC: Regnery Gateway, 1987).

[47] The title of T. Ferris' review of *The Anthropic Cosmological Principle* (Oxford Clarendon Press, 1986) by J. D. Barrow and F. J. Tipler in *The New York Times Book Review* Feb. 16, 1986, p. 20.

[48] A claim of A. H. Guth, reported in *The New York Times* April 14, 1987, p. C1. For other claims of Guth along the same line, see my *God and the Cosmologists*, p. 258.

[49] See Galileo's *Dialogues*, p. 328, 334 and 339.

[50] S. Freud, *Moses and Monotheism* (tr. K. Jones; London: Hogarth Press, 1939), p. 90.

[51] See my Farmington Institute (Oxford) Lectures, *The Purpose of It All* (Washington, DC: Regnery Gateway, 1990), pp. 19-23.

[52] S. Freud, *Civilization and its Discontents,* tr. James Strachey (New York: W. W. Norton, 1961), p. 23.

[53] Spencer did so in his *Social Statics*, published in 1851. For details, see my *The Purpose of It All*, p. 13.

[54] The young man was the son of a friend of mine who until that tragic day found meaning for life in cogitation about extragalactic cousins.

[55] As I have shown in my book, *The Relevance of Physics* (Chicago: University of Chicago Press, 1966, p, 452), the religious commitment of scientists changes in much the same way as does that of other professional groups. Those changes are not, however, reflected in superficial generalizations such as the often referred to materialism of the 19th century. Leading physicists and astronomers (as well as other scientists) of the 19th century professed Christian convictions in a surprisingly high proportion. See on this

the reprinting, with my introduction, of A. Kneller, *Christianity and the Leaders of Modern Science* (Royal Oak, MI: Real View Books, 1995), originally published in German in 1902 and translated into English in 1911.

6

Words: Blocks, Amoebas, or Patches of Fog?

Artificial Intelligence and the Conceptual Foundations of Fuzzy Logic

Words are the fundamental carriers of information. Among words there is, however, one class, the class of numbers, which is very special in one respect: Numbers can be pictured as conceptually corresponding to spatial domains with strict boundaries. This is particularly true of numbers called integers. One can satisfactorily represent the conceptual content of the integer 1 by a right-angled triangle, or by a square. Then the conceptual content of the sum 1 and 1 can again be represented by the juxtaposition of the same geometrical figure. In general, then, the conceptual content of any integer can be represented by a geometrical figure, with strict boundaries. This representational uniqueness of integers may not have been in the mind of of the German mathematician, L. Kronecker, when he told in 1866 the Versammlung der Deutschen Naturforscher

Invited paper read at the Meeting of The International Society for Optical Engineering, 10-12 April 1996, Orlando, Florida. Published in its *Proceedings*. Volume 2716, pp. 138-143. Reprinted with permission.

in Berlin that "God made the integers, all else is the work of man." He most likely had in mind the fact that all mathematics derives from integers. And since, as was noted above, the integer 1 can be represented by a strict geometrical figure, such as a square or a triangle, any other integer can be thought of as the juxtaposition, in the required number, of such a geometrical figure. The result, no matter how large the number, is still strictly circumscribed. Therefore any integer can be spoken of as a strictly circumscribed block, or the sum of such blocks. If the conceptual content of some words resemble blocks, this is certainly true of integers.

This is the reason why basic operations with integers represent the most successful aspect of computer programming. In fact, no matter how difficult a calculation may be, if it can be broken down into summation by finite series, the operation is considered successful. No fuzziness is on hand whenever one can fall back on operations with binary arithmetic. These operations can readily be made equivalent to the physical process whereby the magnetic properties of the units of the chips are given one or the other of their two permissible orientations. These can be expressed in binary arithmetic, which works with integers. Computer programming is a work in terms of rational numbers, that is, numbers that are integers or quotients of integers.

Even irrational numbers are not so irrational as to refuse representation in strict geometrical form. Pythagoras must have known that the length of the hypotenuse of a right angled triangle with equal sides is a very definite length, although it is impossible to express its square root in terms of an integer or of a definite fraction of it. It had to be clear that a definite length, such as the hypotenuse, cannot be the square of a non-definite length, whatever our inability to express in integer numbers the magnitude of that length. The Greeks surmised what two thousand years later Lambert proved, namely, that the ratio of the circumference of a circle to its radius is also an irrational number, although both lengths are very definite.

No geometrical representation can, however, be given to imaginary numbers. The positive square root of -1 is not reducible to a strict geometrical figure, still it plays a very important role in the mathematization of physical reality. This became increasingly realized through the mathematical analysis of alternate currents and other electromagnetic phenomena during the nineteenth century. Since then (it should be enough to think of quantum mechanical theories) the role of imaginary numbers for mathematical physicists has loomed larger and larger. As a result, physical reality appears to have more and more fuzzy edges. This in turn may have promoted interest in what first was called modal logic and more recently fuzzy logic.

This is not to say that this spatial intractability of irrational numbers prompted the erstwhile interest in fuzzy logic and its application. Still it may be that this intractibility remains at the bottom of all problems to which communications and optical engineers address themselves. To cope with that intractability they have two procedures at their disposal. One is the reliance on summation with finite series, or approximation of a quantity in terms of adding any required amount of real numbers, all of which can be imagined as having definite geometrical edges. Implied in this procedure is an arbitrary step, namely, the decision that quantities are ignored beyond a certain cut-off point, which is set arbitrarily. But the same arbitrariness is present in the other procedure as well, or the statistical method. There the cut-off point or cut-off line is merely buried, again arbitrarily, in the haze of probabilities, that again are defined in an arbitrary manner.

But there is a deeper problem with integers, and therefore with all numbers, insofar as they depend on the definition given to integers. The problem in turn reveals that even numbers, or binary counting, to be specific, is not entirely a blissful paradise, where everything is straightforward. For, while the operations with integers may be straightforward, the very starting point is not. For before one performs operations with integers, one must know what an integer is. This however can be achieved only in terms of words that are not numbers. One cannot ponder enough the remark of a great mathematician of the first half of the twentieth century, Hermann Weyl, who warned, in 1951, that operations with numbers depend on non-mathematical words: "One must understand directives given in words on how to handle the symbols and formulae."[1] It would have been even better for Weyl if he had said that one needs words in order to know what an integer is. For unless one is ready to start the march with the second step, numbers cannot be explained in terms of numbers.

Therefore even about mathematics it remains true that "In the beginning was the word." For numbers too can only be understood insofar as they are verbalized. It is here that a kind of fuzziness creeps in, which is deeper than the fuzziness usually dealt with in fuzzy logic. A good ilustration of this is the verbal definition of integers. The definition of integers in any dictionary fully illustrates this. Let us take, for instance, *Webster's II*: "An integer is a member of a set of positive whole numbers (1, 2, 3, ...), negative whole numbers (-1, -2, -3, ...) and zero (0)."[2] The definition consists of 17 words, none of which can be represented by a distinct geometrical figure with no fuzzy edges. Yet the definition seems to be very definite, that is, with precise limits. For this is what the word 'definite' stands for. The three indefinite articles in that definition are by definition the very opposite to something definite. One need not be a linguist to realize that words like 'member,' 'number,' 'set,' 'positive,' 'whole,' and 'negative,' and even the word 'integer', are words with broad meanings. They are useful for the above definition simply because we have an intuitive understanding of integer numbers and we restrict all those meanings in terms of that intuition. The same dictionary gives the meaning of 'and' as "together or along with; as well as." None of these equivalents of 'and' can be given exact geometrical representation, in spite of the spatial touch of the words 'together' and 'along.' No geometrical representation can be given to the verb 'is'. And what about zero? It is, by definition, the opposite to anything that can be delimited and/or represented. Yet, zero is what really defines any number system and plays a most visible part in the decimal system. Still it stands for nothing, which is even less than total fuzziness.

All that fuzziness in the definition of integers can, of course, be eliminated by a simple *fiat*, in the manner of Humpty Dumpty: "When *I* use a word, it means just what I choose it to mean—neither more nor less."[3] Had Humpty Dumpty referred to mathematics, this high-handed manner would have been justified, but only because it works there. But Humpty Dumpty referred to matters entirely free of mathematics. He said, "There is glory for you," by which he meant a "knock-down argument." Such arguments, even if there can be such, presuppose something very different from fuzzy logic. Arguments are a much broader class of rational procedures than are proofs. In fact, the proofs of mathematics are not proofs at all. They are merely identity relations. But this is exactly what definitions of words are not and cannot be, or else all human discourse becomes an exercise in sheer tautology.

For mathematics to work, it has to use, time and again, such knock-down arguments, that is, arbitrary selection of meaning for all its terms. In doing so mathematicians, to use an analogy, simply set an arbitrary limit to the margin of fuzziness they are going to tolerate. This, as I said, works well in mathematics. Elsewhere it does not. Even with respect to words that denote objects very much looking like so many blocks, this arbitrary delimitation of the fuzziness implied in their verbal definition cannot be applied. There is an operational usefulness, at least for bibliographers, in saying that any printed material that contains more than sixty pages is a book. Below that number it is a pamphlet. But if fifty-eight pages are kept together in a nice thick binding, the result will look much more like a book than a pamphlet that has more pages but no binding. Also, nobody would call the Sunday issue of the *New York Times* a book, just because it contains at times close to two hundred pages. It does not help that the size of newspapers is too large compared with typical books. There is a category of books, called elephants, that are bigger in size than our daily newspapers.

The difficulty of drawing a strict line within the margin of fuzziness implied even in the definition of block-like objects can be readily illustrated. Take, for instance, the word, "chest." It is defined as "sturdy box with a lid and often a lock, used esp. for storage." There is a problem with sturdiness. It can be ascribed to a variety of materials, but never to really sturdy materials, such as iron and steel. Sturdiness can also be a variable function of thickness. Then comes the question of size. A smallish box, however sturdy, does not qualify for being called a chest. Or think of the difficulty of setting a strict demarcation line between a low-boy and a high-boy, two pieces of furniture, very similar in many respects.

The problem becomes huge when we try to give a non-fuzzy definition of the object called bench, which is defined in the *Random House Dictionary* as "a long seat for several persons."[4] Of course, it is easy to differentiate a typical bench from a love seat. But constructionally they are the same. One can be gradually transformed into the other. Still one cannot specify the point where a bench becomes a love seat and vice versa. There is a gradual transition, a band of fuzziness, between the two. And what about the fuzziness of love? Let us not, however, go off on a tangent, which incidentally becomes very fuzzy as it turns into an asymptotic line.

Even words denoting blocklike objects have a quality which has something of the most characteristic feature of amoebas that constantly change their shape. The history of the meaning of any word is an example. This is why dictionaries must be constantly rewritten. At times the change is so radical that even a most variable amoeba would not be able to imitate it. Thus the word, "intercourse," which only a hundred years ago still meant exchange of ideas, or conversation in polite society, stands nowadays for an exchange of a very different sort.

When one moves from words that relate to blocklike or simply tangible objects, the geometrical illustration of meaning becomes a hopeless task. Take for instance the word 'information,' which should seem basic to this conference. An almost operational definition of information in *Webster II* reads as follows: "A non-accidental signal used as an input to a computer or a communication system," one of the six definitions of that word given there. The word "non-accidental" alone would warn against trying to give that definition a spatial representation.

There is no operational way of coping with this problem. For the word "non-accidental" evokes the problem, which has not ceased puzzling philosophers of mathematics. The problem is why two negatives make a positive. But this is precisely the problem inherent in the meaning of 'non-accidental.'

The word information evokes still another problem, namely, that it contradicts its apparently visual content, which is suggested by the word 'form' or shape. Yet visualizability is almost totally absent when the meaning of information is given as: "The act of informing or state of being informed." Clearly, to avoid circularity, one has to look up the verb inform, which means, "to inspire or animate with a specific quality or character."

To simplify matters, let us take this definition as consisting of only three words: Inspire, specific, and quality. To render their combined meaning geometrically, we have to rely on the partial superimposition of three meanings. Superimposition means to work with areas. But is the meaning of, say, inspiration, translatable into an area? For even if we use dotted lines to indicate the indefiniteness of the meaning of inspiration, a mere look at the definition of inspiration should make clear the large measure of arbitrariness of the procedure. It is enough to look up that word in the twenty-volume *Dictionary of the English Language*, published by Oxford University Press, to see the real magnitude of the problem. There the various meanings occupy almost a full densely printed page. But even the one-volume *Webster's II* gives six different meanings in seven lines.

Clearly, the only way to illustrate geometrically the meaning of the word information depends not so much on the partial superimposition of areas defined by dotted lines, but on the superimposition of patches of fog. Such a superimposition is foggy or fuzzy indeed. In that respect even an amoeba is much more precise than any word, including the verbal definition of integers. Words, all words, though not the spatial representation of their conceptual content, are best likened to patches of fog. From a distance such patches appear to have sharp edges. Cumulus clouds look like multiple bulges carved out of marble. Even cirrus clouds appear to have edges that, however fluffy, are not fuzzy. But at a close range even the sharp edges of cumulus clouds begin to disappear. No pilot can establish the point where his plane enters a cloud, even a cumulus cloud that appears so well defined from a distance.

Yet, this does not mean that all thinking is fuzzy. Bertrand Russell was very wrong in saying that "all traditional logic habitually assumes that precise symbols are being employed."[5] Traditional logic, like anything else, has had its sensible and not so sensible practitioners. The sensible ones have always remembered Aristotle, who most explicitly stated something very fundamental about numbers. They alone of all the categories do not admit analogous application or interpretation. All other categories are comprised of words, the overwhelming portion of which admit analogous meanings.[6] In other words, in a very profound sense, numbers alone are not fuzzy, even though their verbal definitions cannot be represented by a juxtaposition of strict geometrical figures. The meaning of numbers is definite in a very special sense, which is not true of any other word that belongs to the other categories, such as substance, quality, relation, place, time, position, state, action, or affection. Bertrand Russell was again wrong when he added to his dictum, just quoted: "It [traditional logic] is therefore not applicable to this terrestrial life but only to an imagined celestial existence."[7]

Only the less than sensible practitioners of traditional logic have confused it with reality. If the number of categories, as first given by Aristotle, is correct, reality has at least ten different aspects, of which the quantitative aspect is only one. That quantitative aspect is marvelous to investigate, because it takes only a thinking that moves along the single track of quantities. On that track one need not worry about shades of meaning, all of which relate to analogous uses of the same word. It would hardly be sensible, in fact it would be fallacious, to say that because of the ease of moving up and down on that single track, nothing else can be known reliably. But this fallacy is implied in what Bertrand Russell claimed in saying: "Physics is mathematical not because we know so much about the physical world, but because we know so little: it is only its mathematical [quantitative] properties that we can discover."[8]

Such a claim is self-defeating. Only if we can know properties other than the quantitative is it possible to say that we can know only the quantitative properties. A similar contradiction lies hidden in the statement that all is fuzzy, because this makes sense only if at least one statement—all is fuzzy—is not fuzzy. It is rather sad that prominent logicians like Bertrand Russell and Alfred North White-head failed to note this in making statements that are fondly quoted in the fuzzy-logic literature. The former claimed that "All language is vague."[9] According to the latter, "All truths are half-truths."[10]

Language does not become vague just because the meaning of words, apart from numbers, cannot be given strictly defined spatial representation. In order to define the meaning of any word one has to use half a dozen or more other words. Geometrically speaking, the definition consists in the partial superimposition of a number of other meanings, all of them with indefinite boundaries. As I said before, one may illustrate those indefinite boundaries with dotted lines. Still such lines indicate more definiteness than what actually is on hand. Words may be best likened to the partial superimposition of patches of fog. [See Illustration, p. 87]

To be sure, in using words, the human mind exercises in every instance an arbitrary delimitation of the superimposition. It is in a sense an arbitrary procedure on the part of the human mind. Hence the inevitable demand for ever new definitions that may appear less arbitrary. These, however, cannot by themselves command compliance on the part of any and all. Thus debates will go on forever, even about the most understandable propositions. Not because they are not understandable, but because their understandable meaning is far beyond what can be translated into crisp mathematical or geometrical terms.

While the human mind can cope with this, a communication system cannot. Any software must have definite paths, interconnections, and strictly limited domains representing each and every word, and any operation with words. Hence the operationally unbridgeable gap between human intelligence and the one called artificial. The

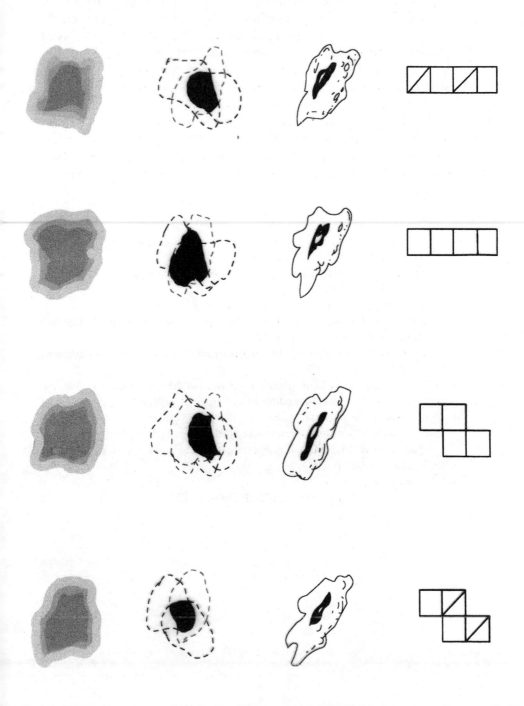

expression, "artificial intelligence," as applied to software or hardware, separately or combined, is indeed very appropriate. Compared with the human mind that can live with the fuzziness of all words, fuzzy logic as an operational procedure is not up to a similar task. For, to mention only one example, what can its statistical method, which assumes ample widths for the same basic meaning, do with disjunctive paradoxical statements, such as the reply which Yogi Berra's son Dale gave to reporters pressing him for a self-comparison with his father. In genuine Yogi Berra fashion, the son replied: "The similarities between me and my father are different."[11] And so is the relation between the human mind and artificial intelligence.

---

[1] H. Weyl, "Knowledge as Unity," in L. Leary, ed., *The Unity of Knowledge* (Garden City, NY: Doubleday, 1955), p. 22.

[2] This and other definitions are taken from *Webster II* (Boston: Houghton Mifflin Company, 1984).

[3] L. Carrol, *Through the Looking Glass,* ch. 6.

[4] For further details, see my *Brain, Mind and Computers* (3rd ed.; Lanham MD: Regnery Gateway, 1989), p. 279

[5] B. Russell, "Vagueness," in *Australasian Journal of Psychology and Philosophy* 1 (1923), p. 84.

[6] This is an implication of Aristotle's statement in the *Categories* (6a) that "the category of quantity does not admit of variation of degree."

[7] B. Russell, "Vagueness," p. 84.

[8] B. Russell, *Philosophy* (New York: Norton, 1927), p. 157.

[9] Quoted in D. McNeill and P. Freiberger, *Fuzzy Logic* (New York: Simon and Schuster, 1993) p. 29.

[10] Quoted ibid., p. 14.

[11] Quoted in *Reader's Digest*, March 1966, p. 170.

7

Beyond Science

I have been asked to talk about what is beyond science. This means that I will have to talk about what is beyond physics. All science is physics at least in the sense that physics has become the pattern for all other sciences. The advent of quantum mechanics turned chemistry into a branch of physics as early as the 1930s. Today, physicists, who think that the standard model of fundamental particles gives the answer to everything, are wont to say that it is now the turn of chemists to do everything physical. The exactness of physics has become the goal of some psychologists, sociologists, and historians. This kind of infatuation with physics has, however, made heavy inroads among philosophers ever since logical positivists raised their illogical heads in the late 1920s. They went very much beyond the claim that science or rather physics was the only respectable subject matter for philosophy. Rather they claimed that philosophy was not

Originally published in W. A. Rusher (ed.), *The Ambiguous Legacy of the Enlightenment* (Sacramento, CA: The Claremont Institute, 1995), pp. 208-13. Reprinted with permission.

respectable unless it emulated the exactness of physics and became thereby a scientific philosophy.

It took but a generation to let that claim reveal its absurdity. The one who admitted this was none other than Professor Ayer, the leader of logical positivists in Great Britain, and the principal headshrinker of a younger generation of English-speaking philosophers. The moment of truth should have come to that establishment when in 1968, in an interview on the BBC, Professor Ayer was asked to say what was wrong, if anything, with logical positivism. "Almost everything was wrong with it," was his answer. If that answer has not yet entered into college textbooks on philosophy, it is only because part of the philosophical establishment staunchly refuses the evidence. Their refusal is so strong as to cause, three years ago, the break-away from the American Philosophical Society of a small group that wanted freedom of thought from the shackles of a philosophy reduced to physics.

Whatever one may think as to what philosophy should be about, it should seem beyond dispute that physics should be about physics. And, in view of the foregoing considerations about the reduction of chemistry and biology to physics, one may also say that if science should be about science, then science should be largely about physics. Therefore the question I have been asked to answer, namely, what is beyond science, boils down to the answer I should give to the question, "what, if anything, is beyond physics?"

To answer this question, one must form a clear idea about the reach or domain of physics or about what is within physics. Without such a clear idea one would place oneself in the nonsensical position of talking about what is beyond the United States, or the earth, or the solar system, or our galaxy, without caring about the limits of our galaxy, of the solar system, of the circumference of the earth, or of the boundaries of the United States. In national and international life the worst kind of tensions arise when doubts, ignorance, let alone wilfulness are allowed about the limits of this or that jurisdiction. The tensions and confusions are no less damaging when the boundaries of this or that science or intellectual enterprise are being treated in a cavalier manner.

The answer to the question, what is beyond physics—which is the other side of the question, what is within physics—can be given in two ways. One is the so-called academic or sophisticated way, in which one's position, even if it is clear, is wrapped in circumlocutions that may have two aims. One is to weaken the critical faculties of the reader or the listener. The other is to satisfy the speaker's or author's vanity. It is far better to be outspoken from the very start. Apart from my bent for it, this latter approach seems alone to be appropriate in this gathering. We have come here to promote clear thought in a culture which, for all its science, is in short supply of that commodity, partly because many so-called humanists are in very short supply of it.

My plain answer to the first question, "what is beyond physics?" is that everything is beyond physics, that is, beyond its competence. My answer to the second question, what is within physics, is that everything is within physics, that is, within its very competence. My lecture has for its task to make sense of this apparently paradoxical if not plainly contradictory pair of answers. On the successful outcome of this effort depends the reasonability of any talk about God, the very subject which you may rightly expect me to broach in this lecture. Do not, however, expect me to rush to that subject without first clearing up some minefields, some hostile, some very well-meaning and therefore even more treacherous, on the road towards that supreme goal of all human reflections.

Moreover, before speaking of God, one should be able to speak of the mind, or soul as a subject beyond physics. Such an appraisal of the mind is today under heavy attack, both rude and sophisticated. Lately, such a rude attack was made by Hans Moravec, senior researcher of robotics at Carnegie-Mellon University, in his book, *Mind's Children*, a book produced by Harvard University Press, of all presses. By "mind's children" Moravec means computers that will automatically generate, as so many children of theirs, further computers far more intelligent than present-day humans. His method is to approach, as he puts it, from the "bottom up" the question of artificial intelligence. He thinks that we have gone sufficiently far with our pliers, wires, and, of course, microchips to be convinced about the reality of machines that already think in an embryonic way. This in turn should assure us, according to him, about the coming, within two or three decades, of thinking machines equal and even superior to the human mind.

About Moravec and his allies, the best one can say is that they seem to be stuck with a bottom very inferior to the grey matter associated with thinking. My excuse for this crude aside is that far cruder remarks have been made about them, and at artificial intelligence conferences at that. One such remark, made by a leader in designing expert-programs, is that the emperor is all crap from the ankle down.

As to the sophisticated method of proving that physics, and physics alone, is competent to deal with the mind or soul, the latest and most widely talked about instance is Professor Penrose's book, *The Emperor's New Mind*, produced by Oxford University Press. The choice of the title is aimed first at what is usually called the strong AI position, which is a variant of the crude AI position described above. The title implies what is very clear from the book's subtitle, namely, that a new physics is the only means of explaining the mind. Professor Penrose does not tell us what that new physics is, except that it has to be a new form of physics based on a notion of randomness, which is also left by him in limbo.

I find his claim useful only for an illustration of the point on which I am trying to focus in this lecture. The point is the overriding importance of having clear notions about the very tool, in this case physics (or science), on which the entire argument of this lecture rests. Professor Penrose's procedure illustrates another, and no less important point for our purposes. He assumes, without first proving it, that physics alone can explain the mind, in fact that physics can explain everything, and that only physics can perform what should seem to be magic. In other words he is guilty of a mishandling of reason for which there was already in classical times a special name, *petitio principii,* or, in plain English, the art of begging the question. That such a technical term was coined more than two thousand years ago and within the very origins of Western culture, may suggest that the mental weakness described by that expression has ever since been plaguing that very same culture.

I would not feel entitled to wash the other camp's dirty linen in public were I not to start with the dirty linen produced by some champions of religion, and the Christian religion in particular, which became a chief ingredient in Western culture. Some theologians have been wont to assume that revelation about this or that was a revelation about everything else. Without proving it first, they went on to demonstrate everything else in terms of that assumption. The error of high-ranking churchmen in the Galileo case derived from their yielding to precisely that kind of temptation. They could only blame themselves for having emulated some Reformers who were the first to damn Copernicus because the earth's motion did not seem to rhyme with their reading of some passages in the Bible. Today, the so-called creationists are guilty of the same *petitio principii*, as they hold for a biblical verity that the Bible is a science textbook. Their grim resolve to dispel the specter of a geological past measured in billions of years is based on begging the point to be proved in the first place.

Having said all this, I may be allowed to say a few words about the dirty linen which non-religionists have produced by cultivating the disease of *petitio principii*. They have often done so by claiming that science or physics can explain everything. Several chapters in my long book *The Relevance of Physics* deal with the spreading of that disease in modern Western culture. Let me therefore recall here but a few false gems that were dumped in enormous quantities and variations on Western culture since the Enlightenment dawned on us. It was during its dawn that the *philosophes* spread darkness by claiming that "all errors of men were errors made in physics." It was during the brave high-noon of the Enlightenment, the mid-nineteenth century, that Herbert Spencer and others took scientific education for a foolproof social cure-all. They bravely predicted on that basis the elimination of all crime within a generation or two. The prediction turned out to be deadly wrong in the literal sense of that word. For better or for worse, this is not the place to recite fearsome statistics about the skyrocketing of crime as man is launching rockets into the very sky.

So much for a historical illustration that it is worth looking carefully at what physics, the paragon of exact sciences, does mean, can mean, and should mean. It is on that look that depends all sanity of talk about science, and, in this age of science, the sanity of any talk about everything else. Here I wish, of course, to speak only for myself. I do so not only for the sake of modesty, which is always commendable, but also with an eye on the intellectual immodesty which sets the tone of our culture. The immodesty in question is a travesty of modesty which lurks behind and beneath almost all references to consensus. Consensus no longer exists insofar as consensus means to agree about some ideas taken for incontrovertible truths, and about some patterns of behavior taken for unconditionally valid norms to which one must consent. The consensus what our modern culture is left with, and indeed what alone tolerates, is to agree not to disagree in order to forestall plain anarchy.

If in the audience there are some who have read my books published during the last ten years, and they amount to more than ten, they should now forgive me for coming up once more with a quote that I have used in print half a dozen times. Yet the quote cannot be repeated often enough or meditated upon at too great length. It originally appeared in *The New York Times*, this notorious promoter of very superficial consensus, with very transparent objectives. In its December 21, 1983 issue (p. A26, cols. 4-5) it carried a letter by W. Pearce Williams, a historian of physics, like myself, a letter he wrote on behalf of Professor Donald Kagan, a colleague of his at Cornell University. What was to be defended was Professor Kagan's decrying some students and faculty there for advocating civil disobedience with a reference to some "higher morality." In his colleague's defense, Professor Williams pointed out that

> What Kagan, I think, was arguing was that there is no "moral" universe to which citizens can now appeal that provides an adequate basis for disobedience to the law. I find it strange that liberals, who insist upon the ultimate relativism of all moral values, suddenly appeal to a "higher" morality (which they are careful not to define) when it suits them. All that went out with the Victorians, and we now inhabit a society in which all moral opinions seem equally valid. . . The point, of course, is, as Kagan clearly stated, that we live in a consensual society in which we often have to do things we don't want to do or even think are wrong, because we have agreed to abide by majority rule. Destroy that argument, and the result is not freedom but anarchy —a condition which the United States seems rapidly approaching.

At this point it would be very tempting, especially for a theologian, indeed a priest, as I am, to take this quote, especially the reference in it to anarchy, for an excuse to go off on a tangent. It would be tempting to delve into that anarchy, the reality of which, at least on the intellectual level, prompted the convocation of this conference. It would be tempting to unmask such possible consequences of that anarchy as social conflagration in a literal sense. (The burning of Washington D. C. may very well repeat itself and indeed spread quickly to other major cities of the Western World.) It would be tempting to speak of those ultimate results which are brimstone and hellfire, of which the first is very much within physics, while the other is, by definition, beyond it. And all this would not be a swindle, though it would be an easy shortcut in the effort to reach the proper goal.

For a start on that road, after all this minesweeping, it may be best to take the phrase, "all that went out with the Victorians," from the foregoing quotation. The Victorians in question were the ones who took Darwin's books for a scientific proof that they could do anything provided the consensus of society was not offended. But again it is not moral duplicity, Victorian or other, which is my target, but the pseudo-scientific underpinning which was found by Victorians in Darwin's *Origin of Species*. That underpinning was Darwin's assertion, which often turned up in the *Origin*, that by relying on the word "chance" he did not mean the absence of cause. Clearly, he should have continually spoken of a cause or causes of mutation, which, of course, were unknown to him, although Mendel had already pointed them out in 1867.

Yet the coming of genetics did not bring about the disappearance of easy references to chance in evolutionary theory. Jacques Monod found it intellectually respectable to take chance and necessity for two different but equally respectable terms of explanation of everything. Even worse proved to be the disease sparked by Heisenberg and Bohr as they claimed that the uncertainty principle definitely discredited the principle of causality. The ultimate form of that disease is the claim that the physicist can literally create matter, indeed entire universes out of nothing. If physics is that powerful there is indeed nothing beyond physics, because even the Creator is bound to be within it. A miserable creator that has to rely on some misguided physicist for the ability to create.

Yet even the most elementary form of causality is not within the competence of physics. No physicist has ever observed causality. Physicists merely observe a succession of events, as all non-physicists do. Only by taking recourse to metaphysics, tacitly or not, can physicists and non-physicists alike take this or that succession of events for causal interaction. This is so, regardless of whether the measurement relating to that succession is absolutely certain or exact, or somewhat inexact or uncertain. The uncertainty principle merely proves that the measurement of conjugate variables cannot be perfectly accurate as long as the physicist has to work with quanta and with non-commutative algebra, the basis of matrix mechanics. No physicist has ever proved that there can be no physics beyond that very useful and very beautiful form of physics known as quantum mechanics.

Only by doing a somersault in logic can the physicist claim that an interaction, which he cannot measure exactly with the tools at his disposal, cannot take place exactly. The somersault is a form of *petitio principii*, or its variant which is a no less elementary error in plain reasoning. Apparently because it was widespread already in classical times, the Greeks labeled it with a phrase, *metabasis eis allo genos*, equivalent to changing horses in midstream, or talking from both corners of the mouth.

Indeed, whenever one talks, whether from one or the other corner of one's mouth, one puts oneself plainly beyond physics, while remaining strictly within it. We are thus back at our initial paradox. In order to come clean with it at long last, I would like to go back to that quotation, or rather to Darwin, who, by the way, cherished the hope that he had laid the foundation of a biology as exact as physics. The first to note or rather to bring this very Darwinian point to the surface was none other than Karl Marx. According to Marx, Darwin provided the biological foundation for a scientific class struggle based on the tools of production. This is not the place to elaborate on that fiasco of historic magnitude, a fiasco best put by President Reagan in June 1989: "The Goliath of totalitarianism has been vanquished by the David of microchip." The fiasco was born in Marxist-Leninist legislation about what the ultimate form of physics ought to be, a form beyond which there could be no real ultimate.

A similar fiasco was hatched by Darwin's insistence on the exclusive explanation of the mind in terms of exact science. He had insisted on this already twenty years before he wrote the *Origin*, an obvious illustration of the belief that it is possible to prove something by merely assuming it. An assumption which is a mere presumption does not fail to exact its due. The moment of truth came to Darwin when he found that in the terms set by him it was easier to speculate about the mind of a dog than about the mind of Newton.

He never had the mental sensitivity needed to appreciate a point far more immediately pertaining to his theory of evolution. The point is not so much that, contrary to the title of his famous book, he never spoke there about the "origin" of species. He merely tried to show the emergence of new species from species already existing. But did species exist? No biologist has ever seen a species. The term species is a generalization, one of those countless terms called "universals," terms that challenged philosophical acumen from the earliest days of Western philosophy. Not, though, the acumen of Darwin, and not even after he had been reminded of this problem by the famous paleontologist at Harvard, Joseph Agassiz, who, though an evolution-ist, refused to join the bandwagon of Darwinism. Those on that bandwagon still have no use for the problem of the reality of species. A fairly recent case is Professor Mayr of Harvard who, in recalling the problem posed by Agassiz, took the view that Agassiz's thinking was warped by his classical education in continental gymnasia.

Being exposed in one's youth to a good dose of Latin and Greek lore has always proved to be a good predisposition for problems of metaphysics, of which a basic one is the ontological status of universals. Any noun—such as tree, stone, house—is a universal, that is, a term that stands for a large number of objects, showing a great variety still within a common property that justifies the use of one and the same term. Still we see only the singular trees stones and houses, and never the universal tree, stone and house. Does this mean that those terms are mere words? For if such is the case, the term species too is a mere word, of whose origin biologists should hand over all talk to linguists. And what about those daring generalizations and inferences on which Darwin relied heavily when speaking about the geological record, the geographic distribution, the homologous organs, atrophied organs, and the like?

The truth of biological evolution, which I firmly hold, rests ultimately on generalizations and inferences that cannot be accounted for unless one is ready to take metaphysics into account. Indeed, in recent years several biologists urged the working out of a metaphysics of evolution. Moreover, long before one takes up the evolution of life, one is faced with a question of metaphysics whenever one registers life. Life is not seen with physical eyes alone unless those eyes are supplemented with the vision of the mind. No biologist contemptuous of metaphysics can claim, if he is consistent, that he has observed life, let alone its evolution.

Now metaphysics, as the very name indicates, is supposed to be about things beyond physics. This is, however, the supposition of idealist metaphysics which, when taken by Kant and Hegel for a guide in physics, produced stupefying horrors and horrendous stupidities. To see this, one should read Kant's *Opus postumum*, a two-volume work still untranslated into English, and Hegel's *Enzyklopedie der Naturwissenschaften*, available in English in three volumes since 1972. The metaphysics I am talking about is the metaphysics of Aristotle who, contrary to the general belief, never wrote a book on metaphysics. The title "metaphysics" was given to one set of his lecture notes only when some hundred years after his death a complete edition of his works was prepared. The title "metaphysics" simply meant that it was the book that followed the one called "physics" in that first edition. Incidentally, even that book on "physics" was on something else, namely, on what today would be called "philosophy of physics."

While in the common estimate, metaphysics is the last step in philosophy, Aristotle called it the very first step. Judging by what he says in the very first of those twelve books, designated now with Greek letters from alpha to lambda, he would have given them the title, first principles. There at the very start of the book alpha, he states that by first principles he means the principles of knowledge that govern reasoning about any subject. There he also states that the theory or science in question is the most intelligible among all theories or types of knowledge or science and that it therefore controls all other sciences or forms of knowledge. He also locates the perennial wellspring of that philosophy of first principles in man's bent on wonderment, or his curiosity:

> For it was their curiosity that first led men to philosophize and that still leads them . . . . For all men begin . . . by being amazed that things are as they are, as puppets are amazing to those who have not yet understood how they work, or the solstices, or the incommensurability of a square's diagonal with the side, for it seems curious that there is something which cannot be measured even with the smallest unit.

Such is a set of illustrations of the wellspring of philosophy, wonderment, which may tempt one to take Aristotle more for an experimental and theoretical physicist rather than for a very outmoded metaphysician, the kind described once by Maxwell: "A metaphysician is nothing but a physicist disarmed of all his weapons,—a disembodied spirit trying . . . to evolve a standard pound out of his own consciousness." The metaphysician Maxwell had in mind was, of course, the Kantian and Hegelian kind which became very fashionable in Great Britain during the last decade of Maxwell's life.

Wonderment is present whenever one generalizes and makes an inference. Metaphysics is present in each and every human word. One does not have to think of angels to see metaphysics on hand. Metaphysics is literally here and now at every moment which we designate by that apparently mysterious word *now*. The word *now* is, of course, mysterious only for those who see mystery whenever science cannot cope with the problem on hand. Such mystery mongerers are the logical positivists and all scientists trapped by them. One of them was, at least on a particular occasion, Einstein himself. On being asked by Carnap, a chief logical positivist, as to what physics can do with the experience of *now*, Einstein replied that it forever escapes the nets of physics and therefore it has no claim to objectivity.

Rarely did a great physicist give away more pathetically the objectivity of science. For unless the experience of *now* is taken for an objective reality, the physicist can never be sure of being conscious of his objective results and cannot communicate them to another conscious being, whose very consciousness rests on the experience of *now*. Einstein's fiasco should seem all the more baffling as he battled positivism even at the price of exposing himself to the indignity of being held by them to be "guilty of the original sin of metaphysics."

Obviously Einstein was haunted by the very vision of metaphysics which Maxwell described and decried in the same context where he spoke of the "den of the metaphysician, strewed with the remains of former explorers, and abhorred by every man of science." Yet, whatever their abhorrence of metaphysics, good or bad, scientists rely on those very words which are so many proofs of the inevitability of metaphysics. They may not, of course, notice that metaphysics is in the wings when the primordial importance of words is noted by a prominent mathematician and theoretical physicist, like Hermann Weyl. He may not, of course, have thought that he was conjuring up metaphysics when he remarked at the bicentenary celebrations of Columbia University, that even in the most abstract forms of mathematics "one must understand directives given in words on how to handle the symbols and the formulae." Does this bring us into remote touch with the dictum, "in the beginning was the word"? In fact, it does.

Yet before going that far, it is best to remember that even numbers are words, that is, generalizations. Any number is taken to be valid if a potentially infinite set of items amounting to quantities is specified by that number. Numbers, which among the mind's products appear least touched by metaphysics, are in fact so many proofs of the inevitability of metaphysics at any step and no matter where one steps in the process of reasoning. It should indeed seem obvious that everything, including the very realm of numbers, is beyond physics.

Here I mean by "everything" each and every thing registered in an intelligent way and even every thought of which one is conscious. Herein lies the answer to the other side of the paradox, namely, that everything comes within the sway of physics. Clearly, material things and their processes have those quantitative aspects which it is the business of physics to ascertain and correlate, or for taxonomical biology to classify, a first step towards quantitative evaluation. In the case of purely material things our knowledge of them consists heavily in a procedure which lends itself to quantitative evaluation.

Even the aesthetic experience yielded by the contemplation of a set of material entities offers itself to some quantitative analysis. Mathematics or geometry does not, of course, explain why the so-called golden cut or proportion is so pleasing to the eye. But even with purely material things it is impossible to pour into a mathematical mould the judgment that they *do* exist. Unless one is an unrepentant Pythagorean, numbers do not cause existence, they rather presuppose it. A telling admission of this is found towards the end of Professor Hawking's *A Brief History of Time*. There he wonders about the difference between existence and his mathematical formulae, but a page later gives up wonderment. No wonder that he concludes that his physics may assure to the universe that very existence for which credit is usually given to the Creator.

The number of aspects that cannot be handled in quantitative terms increases when one considers things that, though still purely material, are living. It is there that one is first confronted with the ineradicable suspicion that living things act, however unconsciously, for a purpose. The realm in which one confronts realities that have even less connection with quantities is, of course, the realm of humanness. While one can argue about the quantity of material goods that turns their removal into an ethical matter, the category of the ethically good and bad cannot be derived from quantitative considerations. Einstein was very much to the point in saying that he had not derived a single ethical value from all his physics.

Yet, even the most spiritual human judgment is possible only insofar as it is part of a neuronal interaction in the brain, a very material entity. Nobody knows whether an exact evaluation of any such brain process will ever be possible, but nobody can say either that this will forever remain impossible. Yet if brain research ever comes up with an accurate mathematical rendering of a brain process connected with a single thought, there remains an enormous disparity between that mathematical formula or function and the thought itself as consciously experienced. It is precisely when considering the brain-mind relationship that we realize that everything belongs to physics yet everything is beyond it.

In the face of this disparity one can have three reactions. The first, that of the materialists, is a mere denial of this disparity between matter and mind or soul. The second, that of the proponents of the two-aspect theory, is a mere registering of those two aspects with no further effort to account for it. The third, that of the dualists, can be of two kinds: one, or the Cartesian, allows the mind only a superficial involvement with the body; the other, or the Thomist, demands a most profound interaction of the two. In both, of course, the positing of a non-material and therefore immortal mind is an act of inference.

Yet inferences are the foundations of knowledge even in exact science. All our discourse about subatomic particles is an exercise in inference. All generalizations and classifications in biology attest the indispensability of inferential methods. Most importantly, human life would not be possible without trust in that very method. The truth of another man's honesty or reliability, and in fact, of his very consciousness, is an inferential truth.

And so is that largest conceivable object of human reflections which is the "everything" taken in the strictest sense. I mean the universe. When I say universe, I mean far more than billions of galaxies. I mean the strict totality of things. Although the universe, taken strictly for that totality, has become, since Einstein's General Relativity, a rational object of science, it can never become an object of observation. Nobody can go outside the universe to observe it. Still the whole modern science of cosmology would be a misnomer were the universe, or cosmos, merely a name.

The universe is the largest and highest object of scientific research. It is a sign of philosophical poverty that our culture has failed to reverberate to the important fact that since Einstein science has acquired an ability to discourse in a contradiction-free manner about the totality of gravitationally interacting things, or the universe. Instead much talk is wasted on the ability of science to penetrate billions of years into the past of the universe, as if astronomers could ever spot the moment of creation. Creation out of nothing, the only kind of creation worth speaking of, is by definition unobservable. The view of Aquinas, that one can know only from revelation that this universe had a beginning, remains unobjectionable from the view-point of science.

Very little has been said about the fact that science has re-established the notion and reality of the universe into that very dignity which Kant had denied it and did so for a specific reason. Kant set as his chief philosophical aim the rational demonstration of Jean-Jacques Rousseau's emotional program, aimed at man's emancipation from all transcendental perspectives and constraints. Kant therefore had to show that it was impossible to demonstrate rationally the existence of that transcendental factor which is God or the Creator. This is why he tried to show in the first two antinomies that the notion of the universe is an unreliable notion. Today Kant would argue the same claim in vain. The vast and exciting science of cosmology is today a plain rebuttal of the most pivotal point of Kant's rationalism. The point is Kant's claim that since the notion of the universe cannot be justified, human reason cannot make a reliable inference to the existence of God, and therefore man remains his own ultimate master, responsible to no one else.

Yet the inference to the existence of God can securely be made once one considers the all-encompassing specificities which science has established about the universe. In focusing on those specificities, such as average mass density, rate of expansion, isotropy and the like, the mind is entitled to ask the question, why is the universe such and not something else. And since those specificities relate to the universe, it is no longer possible to do here what science does in every other context concerning a given set of specificities. In all other contexts any such set is traced to another set. But when the set is the universe itself, this process comes to an end. Beyond the universe there can be no other universe. The idea of several universes is a contradiction in terms. Either those universes are in interaction with one another or not. In the former case they form but one universe. In the latter case they are unknowable to one another.

In postulating a Creator beyond the universe one merely honors the principle of sufficient reason. The postulate demands an inference, which is analogous to the inference that established the existence of the essential spirituality of the human mind. In both inferences one goes to an object about which science cannot say anything. Yet in both cases, and especially in the case when the universe is used as a springboard toward God, science can be enormously supportive.

Here, however, a note of qualification is necessary. While it is true that the universe is a proper object of scientific reflection—the entire field of modern scientific cosmology is a proof, if proof were needed—still science is not the source of the notion of the universe and of man's intellectual trust in its reality. That source is man's philosophical capacity whereby he can form the so-called universals, the very basis of his concept formation and ability to speak in words. Modern philosophers who have poured scorn on the question of universals have been forced, time and again, by the force of logic to treat scientific cosmology with contempt. If one is to look for the ultimate reason of Whitehead's and Karl Popper's distrust of modern scientific cosmology, one has to look in that direction. It should be no surprise either that both held the universe to be capable of taking on all conceivable sets of specificities during its never-ending life-span.

To postulate eternity and omnifacetedness for the universe is not science, not even philosophy, but plain mystery-mongering. It belongs to the class of statements which institutionalized materialism forced on its scientist victims. I could feel only pity for one such victim, Prof. Ambartsumian, when as a member of the cosmology panel at the 1978 Düsseldorf World Congress of Philosophy he claimed that the most evident result of empirical science is the demonstration of the eternity of matter. As a member of the panel I knew that he had arrived in the last minute to Düsseldorf—it was still the Brezhnev era —in the company of two KGB agents. This is why I did not press Prof. Ambartsumian beyond the point of noting that an experimental proof of the eternity of matter would demand an experiment that would go on for eternity. Otherwise, the materialism he was forced to support would be a call for a metaphysics which it is supposed to discredit.

The road I tried to trace out in the effort of going beyond science is obviously a philosophical, indeed, a metaphysical road. For the overwhelming part of that road physics and metaphysics are like two sides of the same coin. It is only at the very end that this coupling of science and metaphysics is used as a springboard to allow the mind to raise its eyes higher to a realm which is strictly the realm of the spirit. No recourse during that road was made to the inability of science to cope with this or that problem as if any such inability would entitle one to have a direct recourse to God. What science cannot do today may well do tomorrow or the day after. Any quantitatively or empirically stated problem can have its answer in science even if that science is not yet known. In that sense there is nothing beyond science. One is, however, pushed beyond science as soon as one faces basic questions about existence and knowledge. One indeed has to go beyond science in order to have science itself.

The proof of this cannot be scientific, partly because there are no proofs in science. In science we have only identity propositions, marked with that equality or equation sign, which cannot be dispensed with even in the most esoteric forms of theoretical physics. It is precisely that equation sign which cannot play a part, say, in the recognition of any reality in front of us. The relation between the knower and the thing known is not equality. Knowledge is not an exercise in democracy. It is rather the finest form of rule over things known. Those who, for a false modesty, refuse to recognize that they rule when they know, can only end, as intellectual history often shows, in skepticism whenever they do not balance that false modesty of theirs with an equally disreputable recourse to inconsistency in reasoning.

Those who are not afraid of that role of ruling by thought over things and over anything, have a unique opportunity, indeed an obligation, provided they love consistency. They can go so far beyond anything, including science, and indeed all things, which is the universe, as to recognize the One who rules over all and made thereby possible that science which can find a rule and a law everywhere in the physical universe. When man prays, "Your will be done on earth as it is in heaven," he is asking for the grace to behave on this earth as lawfully as do all heavenly bodies of which, as science taught man, the earth is but one. It was through that lesson that science put, in the long run, the universe into man's grasp so that he may grasp a Being infinitely beyond whatever science can ever show him.

8

The Reality of the Universe

In this age of scientific cosmology no topic may seem more superfluous than the reality of the universe. On the contrary, the topic is not only vital but also greatly neglected. Surprising as this may seem, the universe has failed to be given its due not only in philosophy and theology but also in science.

In science or in scientific cosmology the status of the universe has become, through studied neglect, outright tragicomical. The comedy is in full view when one hears a prominent scientist claim that he can create entire universes literally out of nothing. This merely proves that he has greatly neglected to study the terms universe, nothing, as well as the term to create, to say nothing of the word "literally." Similarly comic is the view, clothed as it may be in complicated mathematics, that there are as many universes as there

This chapter was first delivered in Italian at a symposium at the Lateran University, Rome, 1992 and published as "La realtà dell'universo" in M. Sorondo (ed.), *Physica, Cosmologia, Naturphilosophie: Nuovi Approcci* (Roma: Herder/Università Lateranense, 1993), pp. 328-41.

are scientific observers. One need not be a mathematician to spot something farcical in the claim that some acts of observation, perhapsa mere thought, can make the wave function of the universe "collapse" into a reality of cosmic dimensions. What one actually witnesses in these cases is the collapse of reason itself.

Tragedy lurks, however, between the lines of scientific books that are about a so-called conscious universe. The tragedy is of modern man, unsure and perhaps afraid of his conscious self—a self ruined by critical philosopy, psychoanalysis, cultural relativism, and the glorification of randomness. Overawed by his science, modern man now endows with consciousness the universe by transferring to it his own consciousness and perhaps his responsibilities as well. In the process he forgets that he can know the universe only inasmuch as he is genuinely conscious of things external and, through his knowledge of them, of himself.

The seeds of this macabre outcome were sufficiently clear for those willing to see from the very moment when modern scientific cosmology was born. That birth was marked by the publication in 1917 of the fifth or last memoir of Einstein on general relativity. There, for the first time, non-Euclidean geometry was systematically used to obtain a quantitative account of the totality of gravitationally interacting things. There was much more to that memoir than meets the eye, especially the eye that can see matters more than purely scientific.

The scientific content of Einstein's memoir can be summed up in a few lines. First, there is the assumption that the distribution of matter in the universe is homogeneous on the large scale. This means that by counting the number of galaxies in a fairly large volume of space, one can obtain a reliable figure for the average density of matter throughout the universe. The second assumption is that permissible paths of motion for gravitating bodies correspond to lines that can be drawn in spherical Riemannian geometry. The one line that cannot be drawn in such a geometry is a strictly straight infinite line. This means that all gravitational lines of force have a curvature. Therefore no mass, whatever may be its initial velocity, whose upper limit is the speed of light, can escape into infinity. Consequently, the universe can be taken for a finite sphere. The radius of that sphere corresponds to the permissible path of motion with the minimum of curvature. The total mass within that sphere can be calculated from the average density as defined above.

This picture of the universe is scientific only in a restricted sense. Unlike other properly scientific propositions, this picture cannot be subjected to direct verification. No observer, no instrument can be taken outside the universe to observe it and test experimentally either its total mass or its radius. The indirect supports of that picture are, however, considerable. First, it is possible to have experimental proof of the bending of the permissible paths of motion. It should be enough to think of the measurement of the bending of starlight around the sun. Similar evidence may be obtained by using entire galaxies as gravitational lenses. While this bending of the light is also predicted by the Newtonian theory of gravitation which assumes a flat infinite Euclidean space, the value predicted is only half of the true value obtained and observed on the basis of Einstein's theory.

A further support of the Einsteinian view of the universe is that it is applicable to the overall dynamics of the universe or its expansion. This dynamics can be followed, in theory, in the reverse direction to stages where the total mass of the universe is contained within exceedingly small volumes, much smaller than a pinhead.

Great as such scientific successes are, they do not guarantee a grasp of the strict totality of material things, a totality which is the universe properly so-called. The reality of the universe is not necessarily equivalent to the total mass that can be computed on the basis of Einstein's fifth memoir or of any of its many refinements that today constitute a most intensely studied branch of physics. First, progress in observational astronomy has not failed to spot strange new objects, such as neutron stars and quasars with enormous masses. Black holes, if unmistakeably observed, to say nothing of heavy neutrinos, would also necessitate a revision of the average mass density. The homogeneity of the large-scale distribution of mass may also need considerable revision as shown by recent data (and disputes) about the formation by galaxies of a great wall at about five billion light years. The dynamics of galaxies shows departures from Kepler's laws which could be explained if the density of matter were at least ten times or perhaps 100 times higher than the one actually computed.

These problems affect scientific cosmology only in details. They do not touch on the basic assumption of Einstein that cosmology has to rely on non-Euclidean geometry and that the average density of matter is a clue to its total amount. Two points must, however, be emphasized. One is that the total amount is not necessarily the material taken in its strict totality. The other is an expectation now three quarters of a century old. Novel and baffling as may be many findings concerning matter in cosmic spaces, they did not necessitate an abandonment of the perspective and method of cosmology as outlined by Einstein in 1917. This is precisely the point which was not taken into account in Margaret Geller's remark: "I often ask myself what we will learn about the large-scale structure during my lifetime. There will be surprises, answers to old questions and the uncovering of new puzzles. At every stage we will think we understand, but at every stage there will be nagging doubts in the minds of those who wonder."

The covert agnosticism which reverberates through these words is not their sole notable feature. Leading cosmologists seem to be just as unappreciative of an important aspect of Einstein's work as he himself was. Neither in that memoir of his nor in his subsequent writings did Einstein make much of the fact that immediately prior to him the science of cosmology was in a schizophrenic state. For several decades before 1917, it had been generally held that the observable universe was limited to the Milky Way which at that time was believed to be much larger than other spiral galaxies. Beyond that 'observable' universe there lay, so it was believed, an infinite expanse of matter, all uniformly distributed in form of stars or galaxies. Incidentally, during this period there was at least one attempt, which may not have been unknown to Einstein, to treat the Milky Way, which at that time seemed to included many spiral galaxies, in terms of Riemannian geometry.

At any rate, the idea of an infinite Euclidean universe should have been considered with the greatest suspicion. Already in Newton's time it was pointed out that such a universe implies two contradictory or paradoxical features. The less serious of these is the optical paradox, namely, that in such a universe the total intensity of starlight should be infinite at any point. This paradox is valid only if one attributes an infinite duration to bright stars. Very serious, however, is the gravitational paradox or the consequence that in an infinite Euclidean universe the gravitational potential is infinite. A universe with such a paradox is not a physical possibility.

There is still another very serious difficulty in the idea of an infinite Euclidean universe. The difficulty is that it includes by definition an infinite number of objects. Such a number, an actually realized infinite quantity, is, however, a mathematical impossibility. The difficulty was noted, as far as I know, only twice in the entire pre-Einteinian cosmological literature. First by Halley, who brushed it aside; then by Lambert who held it to be a decisive argument against the infinity of the universe. Lambert was also the one who worked out for the first time in the history of scientific cosmology a contradiction-or paradox-free account of the universe. It consisted in a hierarchical grouping of galaxies all of which revolved around a central body. The centrifugal and centripetal forces were in that universe perfectly balanced, according to Lambert, who failed to take note of the possibility, however infinitesimal, of resistance in cosmic spaces. Consequently, his world-model could not be stable.

These historical antecedents may help one appreciate the magnitude of Einstein's achievement. With him the universe, under certain conditions, became a genuine object of scientific inquiry. While he was not entirely oblivious to this point, he failed to appreciate, as did most other cosmologists after him, the philosophical background of his work. This is particularly curious on the part of Einstein, who claimed to have studied Kant's *Critique of Pure Reason*. Einstein did not remember that Kant's pivotal target was neither reason, nor God, but the universe itself. The universe, to recall Kant's major contention about it, was the illegitimate product of the metaphysical cravings of the intellect. Kant tried to show this by an apparently scientific argument, namely, that science could not establish whether the universe was finite or infinite. If this was true, one could also argue that the universe was not a reliable stepping-stone to God. Such was Kant's strategy to provide the support of 'pure reason' on behalf of Rousseau's emotional claim that man was his own master.

If this was true, belief in God was smashed, a point concisely put by Moses Mendelssohn when he called Kant "the one who shatters all." Actually, he should have called Kant the great tranquilizer. Kant merely provided a most dubious intellectual argument for Western man. That man had been looking for some time for a tranquilizing pill of a more respectable kind than that provided by David Hume's self-destructive scepticism. Kant delivered the goods in a package which, because of its convoluted style, appeared learnedness itself. With that huge pill lying half-digested in his stomach, Western man could lull himself into the belief that pure or critical reason and God had nothing to do with one another. Therein lay the intellectual justification of the real issue, namely, that man could rule his life with no transcendental constraints in his way.

Since God is at safe remove from sophisms, Kantian or other, and because the *anima naturaliter Christiana* is also indestructible, Kant could only destroy man's rational confidence in the existence of the universe. Indeed, in post-Kantian philosophy the universe became a mere word, if it was still used at all. The word "universe" had no place in the positivism of Auguste Comte and of J. S. Mill. The pluralistic universe of W. James's pragmatism is just as illogical as is the universe within Dewey's empiricism. There is no room for the universe in existentialism and in phenomenology. Whatever else the universe may be, it is not a phenomenon. And insofar as the universe means coherence, it is irreconcilable with the radical separation among all events as postulated by Sartre. It should be no surprise that deconstructionists can at most repeat Kant's false arguments against the universe. As if to prove how far they are behind the times, they are the least willing to construct a philosophy about the universe in this age of scientific cosmology.

The universe is at best a mere concept in various forms of idealism, Berkeleyan or Hegelian. Trust in the reality of the universe could but suffer through the fondness of many modern cosmologists for idealism. Einstein's great memoir was hardly a dozen years old when De Sitter declared the universe to be a mere mathematical formula. As always happens in the history of thought, bad philosophy seeks a remedy in something equally dubious. Thus it became quite fashionable among modern scientific cosmologists to think that some very simple mathematics about the universe must turn into its physical reality.

This wishful thinking has come into vogue through Hawking's *Brief History of Time*. There he tries to make it appear that a truly simply mathematical formalism about the universe makes the Creator unnecessary. Hawking does not seem to realize that, pleasing is it may be, Platonism just does not work. It can only turn some heads that were not entirely sober to start with. (True to their names, Plato's symposia included a good deal of wine-sipping on comfortable couches, the forerunners of plush academic chairs). Only such heads would see redeeming value in Hawking's suggestion that in order to make modern scientific cosmology fully understandable to modern man, one must have a recourse to the insights of such great philosophers as Aristotle and Kant. Elementary philosophical insights must be lacking whenever such a recommendation is made, which is tantamount to entrusting the cabbage to a goat.

For Aristotle the universe was, of course, a reality, coherent and consistent, and it remained such throughout the entire Aristotelian tradition. Yet neither in Aristotle, nor in the Christian or Scholastic version of hylemorphism, do we find a genuine philosophical proof that there is a universe. Representatives of both schools could claim as an excuse that no proof was needed concerning the reality of something, the cosmos, whose very confines, the sphere of the fixed stars, were visible to the naked eye. Furthermore, instead of proving that there is a universe, Aristotle declares in an a priori manner that there has to be a universe as a sort of ultimate being and that it has to be spherical, imperishable in its upper regions, whereas its lower regions must be made up of four perishable elements.

Thomas, the theologian, proves that there has to be a universe on the ground that God's work must show the kind of rationality which is coherence. Thomas, the philosopher, does not, of course, admit Aristotle's claim about the necessity of the universe. Thomas held the idea of a geocentric universe to be merely most reasonable. Thomas, the philosopher, offers nowhere a strict philosophical argument about the reality of the universe. This is certainly curious in view of Thomas' five ways of demonstrating the existence of God. In all of them the idea of the universe runs between the lines but is not outlined at all. This is especially true of the third way, which is often taken for a cosmological argument. That contingent entities do not explain themselves by a regress to infinity is an argument which implies that they form a finite series in a metaphysical sense. Tellingly, Thomas also holds that an actually realized infinite multitude or quantity is impossible.

Scholastic philosophers have, of course, followed Thomas in discussing the question whether the universe is one or many. Scholastic textbooks on cosmology offer many illustrations of this. But to prove that the universe is one may seem to beg the question whether there is a universe. In fact, time and again the proof offered becomes an unfolding of the notion of the universe and of the contradictory consequences of assuming several universes. In this case one may simply ask whether those universes interact or not. If they do, they form only one universe. If not, only one of them can become the object of a rational discourse which is not severed from empirical reality.

In fact, Christian Wolff, the first to write a book with the word cosmology in its title, may have, by proving the reality of the universe from its definition, discouraged efforts to ask the question whether there is a universe. This is perhaps the reason why Protestant and Catholic imitators of Wolff—and they were many—did not care to raise, let alone answer, Wolff's question. In fact, to speak only of Catholic authors of books on cosmology, almost half of them fail to discuss the universe at all. The other half take up the universe only to prove its unity. Very recently the situation was made worse by attention being shifted to the evolutionary aspects of the universe, as if it were possible to prove that it was permeated by an upward *élan vital*. What happened was to hang Christian vocabulary on Bergson's and Whitehead's pananimistic and pantheistic flights of fancy. For both, the universe is the successive realization of all possible forms of existence. That this proposition, patently non falsifiable, is also subscribed to by Popper, shows once more that inordinate keenness on logic often invites rank illogicalities.

Clearly, in this empirical or scientific age, where coffee-table books about the cosmos have become a dime a dozen, a more reliable approach to the universe is needed. The reason for this should be obvious from the Christian viewpoint. That viewpoint is codified in all the Creeds. They all start with a variant of professing faith in the Father Almighty, maker of heaven and earth, of all things visible and invisible. This tenet of the Creed has more interest to it than the fact that it is a tenet which can be known both by reason and by faith. A still unexplored interest of that tenet lies precisely in the direction of our topic, the reality of the universe. The Creeds nowhere ask us to believe in the reality of the universe. Rather, the Creeds assume the reality of the universe and challenge us to believe in its maker, the Father Almighty.

The very word "almighty" when taken in its Greek original *pantokrator*, proves precisely this. It includes the *pan* or *to pan*, a chief expression in classical Greek for the universe. The universe is denoted in the phrase, "heaven and earth," of Hebrew origin. More philosophical is the expression, "all things visible and invisible," the third way in the Creeds of calling to mind the reality of the universe. The least one can say is that the universe is taken very seriously in the Creeds.

And rightly so. The whole rationality of faith in the supernatural —the Fall, Incarnation, Redemption, Resurrection, Final Judgment,

New Heaven and New Earth—depends on that first tenet of the Creed about the Father *all*-mighty. This tenet becomes void of meaning if the universe, *the all*, cannot be shown to be real. Kant's strategy— a strategy not so much philosophical as ideological—should now stand out in its stark vividness. Any doubt cast on the reality of the universe will be a long shadow cast on the rational recognition of the existence of God and consequently on the fact of revelation.

How to prove then that there is a universe, a strictly coherent totality of things? The very word universe may provide a clue. The word universe is a variant on universal, this most basic and most abused term of philosophy. All our words are universals, that is, they denote a class, a group of things, insofar as they have common characteristics. What is common is not, however, singular. The very process of understanding means precisely the sighting of the universal in the singular. It is also well known that universal characteristics can easily be grouped in one of the ten Aristotelian categories. Equally well known is Kant's handling of them. With him the categories become a priori forms of the intellect through which all sensory impressions are filtered into intelligible terms.

Since space as experienced by the senses is three-dimensional, Kant had to hold the Euclidean perception of space to be a necessary category. He then should have stated that the universe had to be infinite in the Euclidean sense. It remained to some of Kant's 19th-century admirers to draw this inference. One of them was Olbers, who in 1823 called attention to the optical paradox of the infinite Newtonian universe. His famous essay began with a reference to Kant as the one who had demonstrated the infinity of the universe as the only kind of universe worthy of God. Whether this reconstruction of Kant's thought was accurate or not is irrelevant. The fact is that by 1823 Kant could be invoked as the supreme seal on the infinity of the universe. This infinity certainly followed from premises laid down by Kant. The idea of an infinite Newtonian universe should be credited to Kant, if the idea deserved any credit at all. The idea found indeed wide currency with the rise of Neokantianism in the second half of the 19th century. The Hegelian-left, which should be best called the Kantian-left, concurred with its celebration of an infinite eternal material universe as the ultimate entity, a dogma of a now defunct Marxist orthodoxy.

As is well known, Kant had to deny rational status to the universe partly because it could not be the object of sensory perception. Had Kant been satisfied with saying that the universe is not a phenomenon, he might have performed a useful service in philosophy. Scholastic philosophers might have then been aroused from their cosmological slumber, especially after Leo XIII's *Aeterni patris*. Efforts might have been made to prove that there was a universe.

In fact, a good argument could have been worked out by starting from a premise, the universals, which Kant successfully expropriated to his own purposes. One needed only to have paid attention to a little remembered phrase in Thomas Aquinas' commentary on Aristotle's *De anima*. There Thomas emphatically states that even the categories of the mind are abstracted from sensory experience. If anything is alien to Thomas' thinking it is the a priori.

Once this starting point is taken with Thomas, it is possible to avoid the trap set by Kant and show that there is a universe. Incidentally, not only the notions of God, universe, and soul could be caught in that trap but that of science as well, a point of particular importance in this age of science. If the essence of understanding is to seize on the universal in the particular, then to understand is to grasp some totality. The totality as such is never seen. The idea of a chair is not open to sensory perception, although the senses can unfailingly register the object called chair.

This disparity between the universal idea and the particular thing can especially become a thorn in the side of the biologist hostile to metaphysics. Such a biologist is the Darwinist who at best can pretend that no philosophical problem is presented by such universals as are all species, to say nothing of higher levels of universals needed by him, such as genera, orders, classes, phyla, and kingdoms. But if he keeps his eyes open to the metaphysical strain hidden in those biological universals, crowned by kingdoms, he may even qualify for the blessing of the words: "You are not far from the Kingdom of God," that ultimate form of all created totalities.

But to speak rationally of that supreme and supernatural totality one must first gain hold of that totality which is the universe. To speak of a "new heaven and new earth" makes sense only if there is an assurance of a mere "heaven and earth" or the universe. For this purpose one may use a variant of Thomas's fourth way of proving the existence of God, a way based on the grades observed within any class of perfections. The argument then begins by stating that any totality, as a form of perfection, is really and consistently understood only insofar as it is set against a larger, more inclusive totality. But this again is subject to the same restriction. Something analogous to Gödel's theorems is implied here though without the precision as well as the poverty of mathematics: any totality presupposes for its consistent understanding a larger, a more inclusive totality. As a result one may conclude that the sensory understanding or grasp of any totality depends ultimately on the reality of its supreme kind, which is the universe. Only this way can regress to infinity be avoided.

Such is an argument which, reliable as it may be, may not have a convincing value in an age that has only suspicion for metaphysics. Since, rightly or wrongly, this suspicion is fomented by science, one may ask whether there is a proof of the reality of the universe with a distinctly scientific flavor. A scientific flavor means that only the addition or coating need be scientific, not the substance or the crucial starting point. Science in fact is unable to assert even the reality of its instruments, although scientific work has to start with them. Scientists must presuppose the reality of matter before they can talk of its quantitative properties.

Therefore, starting with real matter, whose totality the material universe ought to be, we take a metaphysical ground. This means that subsequent use of metaphysics may not be frowned upon as an unwarranted instrusion. A universal characteristic of real matter is that it can be counted or measured. The basis of all measurement is the set of integers which exclude only one unit, the unit of infinity. An actually realized infinite quantity is a contradiction in terms. Whatever one may assert about Cantor's transfinite numbers, one is wise to recall Hilbert's warning that they have nothing to do with physical reality.

From this consideration of integers it follows that real matter can exist only in a finite quantity. This conclusion is not destroyed by the fact that in some non-Euclidean geometry (such as hyperbolic space) the gravitational paradox does not arise even if the total amount of matter is infinite. This is only so because in such a space the idea of truly central forces is meaningless. But even in such spaces an actually realized infinite quantity remains an impossibility.

Mathematics is therefore very helpful in justifying the notion of a universe which is a totality because of the limitedness of matter to a finite quantity. But this totality becomes a universe only if it embodies genuine coherence or a common ground of understanding.

In this aspect, too, science or mathematics should seem very helpful. As I said before, integers provide the basis for the kind of universal understanding which is measurability. The physical universe is assumed to be measurable throughout. Any doubt cast on this assumption undermines the universal validity of scientific laws. Mutually irreducible as some of those laws or forces may appear, they have one thing in common which is the casting of all of them in quantitative terms. This is why one may say that the gravitational force is weaker than the electromagnetic force by a specific quantitative ratio and that electromagnetism is again weaker than the nuclear force by a ratio no less specific. In fact, some leading scientists try to find in purely numerical ratios the basic structure of the material universe.

That idea that integers assure universal coherence received a striking proof in the construction of electronic computers. They all are based on coding sundry data in the binary number system. Trust in cosmic coherence as carried by integers was demonstrated when a mathematical message set in binary code was engraved on a metal plaque carried by Pioneer X into interstellar spaces.

The phrase, "God made the integers, everything else is the work of man," contains therefore vistas far deeper than that of a striking contrast. Trust in quantitites reached new heights in antiquity with the coining of the biblical phrase, "God disposed everything according to measure, number and weight." This was no mere praise of numbers, such as the Pythagorean incantation: "Bless us divine number, thou who generatest gods and men!" Plato's more philosophical encomiums of numbers remained hollow declarations. He was unable to solve the problem he registered in *Phylebus,* which he brought to a close with a reference to the invariable intermingling of the one with the many. As shown by the history of philosophy, this problem remains intractable except within the framework of Christian philosophy.

An indirect proof of this is that philosophers, who confined the horizons of reason to this world, and became thereby "worldly" in that most profound sense, lost out on the world or the universe itself. They did so by intent. They sensed in terms of their Humean and Kantian heritage that if they accepted the universe they could not claim that there was nothing to be seen beyond it. Bertrand Russell contradicted that entire heritage when he stated, in his memorable radio debate with Father Copleston, that the universe is a brute fact with nothing to be seen beyond it. The very word, nothing, this most metaphysical creation of the human mind, proved the very opposite. For if every insight is restricted to the sensory, the very denial of all sensory, indeed of all existence, is impossible to account for. Einstein, too, failed to realize the measure in which he parted with all "worldly" philosophers when he spoke in the same vein at about the same time as did Bertrand Russell. The parallel is even closer. While Bertrand Russell was confronted with a priest, Einstein felt that he conjured up the reality of priests. "I know," he wrote to his friend Marcel Solovine, "that priests would make much use of my achievements. There is no remedy to that. Let the devil take care of them."

Priests, theologians, Christian philosophers, and Christian intellectuals in general would do well to take real care of the universe, the very foundation of their faith. To have an intellectually respectable hold on the universe demands more than shallow exercises in rhetoric. These do not gain any depth by being sprinkled with the holy water of quasi-theological terms. Christian philosophy has yet to do justice to the universe. Fondness for quantities first became a cultural climate during the age of faith. Being an age of faith, it possessed undisturbed the reality of the universe.

In all likelihood that age will not come back. Today we are removed a great distance from it. In the Middle Ages it was natural to put the emphasis on the words, "according to measure, number, and weight," that is, on the quantitative properties of every thing. Today, after four centuries of science, and after as many centuries of non-Christian philosophizing, the emphasis needs to be put on "everything," or the universe. This we should not do in order to appear fashionable. Our duty lies in the fact that eternal truths ought to be proclaimed in a manner adapted to the particular needs and receptivity of times. Nothing serves better the dictates of our duty than to hold high the timeless truth of the reality of the universe and through it the truth about its real Maker.

9

A Telltale Meteor

Its size is a potato, its color first looked green, though not because it had first been handled by little green people from outer space. Its scientific name is ALH84001, which is much less mysterious than it appears to be. ALH stands for Allan Hill, an area in the Antarctic, 84 for 1984, or the year when the meteor was found there; 001 tells that it was the first to be found and labeled among a dozen similar meteors. It looked green until it was removed from the glare of the Antarctic ice.

Had that meteor not aroused, about two years ago and almost by chance, the interest of Dr David McKay, a geologist with NASA, the meteor would not have produced a new outburst of interest in extraterrestrial intelligence. For a few days after Tuesday, August 6, it was the hottest story. Newspapers will, however, drop it soon like a not so hot potato, because even scientific miracles last only three days. Who is excited today about the discovery of penicillin, of

First published in *Wanderer,* August 22, 1996, p. 5. Reprinted with permission.

antiparticles, of the double helix structure of DNA, or even about the landing on the moon?

Indeed just about three days after the outburst of euphoria, *The New York Times* warned the readers of its Sunday, August 11, issue that "the latest hypothesis put forward by NASA extends a tradition marked chiefly by disappointment." Reports about canals on Mars proved to be a work of imagination. The same illustrious paper did not add that in 1964 a blue-ribbon committee of the National Academy of Sciences had assured President Johnson that NASA would certainly find on Mars such lower forms of vegetation as moss and grass. In 1976 the Viking space-probe found no life on Mars, not even traces of death. In the Op-ed page Stephen Jay Gould was allowed, under the heading, "Life on Mars? So What?" to scoff: "A hypothetical argument for the probable existence of Martian fossils is scarcely worth the effort of an E-mail message." For even a direct evidence of bacterial life on Mars would merely "universalize the Age of Bacteria; humans remain as gloriously accidental as ever."

Now if ALH84001 leaves humans on earth "gloriously accidental," extremely improbable should seem the existence of human-like intelligent life elsewhere. Yet this improbability received all sorts of benevolent evaluation by the media when that potato-size meteorite caught, if not world-wide, at least national attention. Tellingly, *The New York Times* found nothing to criticize in the high-handed action of the editor of *Science* which in its August 16 issue carried the official report. While Dr McKay used the word "possibly" at a crucial juncture, it was blue-penciled by the editor.

The word "possibly" is most appropriate to use in reaching a conclusion about the analysis of ALH84001. The meteor contains (1) PAH (polycyclic aromatic hydrocarbon) molecules, (2) magnetites, (3) iron-sulfate, (4) and carbonate globules. The simultaneous occurrence of these four items strongly suggest bacterial activity. The fact these items are more abundant toward the meteor's central parts greatly reduces the possibility that they are due to contamination on earth.

Less strong is the meteor's connection with Mars. The Viking space-probe carried with it instruments to analyze the Martian surface and radioed the data back to the Earth. Quite a difference from our way of knowing the Moon's surface, which is based on extensive analysis of a large number of rocks brought back from the Moon itself.

The age of the meteor can be evaluated with great precision, the data giving about 3.6 billion years. One can also compute the minimum time during which the meteor had to be on Earth. This is about 13,000 years or so. It is reasonable to see the triggering of the mechanism whereby the meteor came to the Earth in the grazing impact of a huge comet on Mars about 15 million years ago. Much of the debris detached from Mars went into orbit around the Sun and some of it was eventually captured by the Earth.

A strong as well as weak point in the argument that sees in the meteor the carrier of traces of life from Mars relates to worm-like images, or elongated, segmented forms. Are they truly traces of cell-like structures? That question can be answered positively only when traces of cell-walls are found in that meteor.

The meteor has become the hottest news precisely because of sanguine expectations that such will indeed be the case. For decades now the following two points have become almost an article of faith in our scientifically conditioned culture. One is that life arises spontaneously everywhere in the universe where proper chemical conditions obtain. Second, once self-duplicating units, however primitive, are on hand, life is irresistibly on its march toward ever higher forms, even to forms that possess intelligence, and presumably in much higher forms than is needed for the development of science as we have it on the Earth now. If, however, this is true—so the argument goes—man is just another happenstance formation among countless others in the incredibly vast universe.

This last inference was not spelled out in such a drastic way when the meteor held the center stage of attention. But enough was said about "important philosophical consequences" and "exciting possibilities" to suggest the all-important point: Religion may be cast overboard once and for all if the meteor lives up to its billing, a billing much more ideological than scientific. Indeed in the couple of interviews which I reluctantly gave to various newspapers on Thursday, August 8, almost immediately there arose the question: What does this mean to religion? What about the Bible?

I said, reluctantly, because I have already had a long and frustrating experience of the reluctance of reporters in general and of science writers in particular to ponder important distinctions.

First, what or which religion is to be discussed? Speaking of the Roman Catholic faith alone, one can point out that the Church has no special tenet about the origin of purely organic or biological life.

About the origin of the human mind or soul the Church has, of course, a very distinct teaching: Each and every human being is from the moment of its conception invested by the Creator with a strictly non-material or immortal soul, an entity made in the image of God himself.

Of course, about the second point nothing can be argued scientifically. Scientific argument hinges on experimental verification. Nothing can be experimented about the human soul which is chiefly evident through human consciousness, pivoted on the experience of the *now*. Once Rudolf Carnap, a leader among positivist philosophers, asked Einstein what can science do with that experience. Einstein replied that since physics cannot deal with the *now*, consciousness should be considered a purely subjective factor. Neither he nor Carnap perceived that all criminal law collapses if society accepts that "scientific" view of human consciousness. The same consequence arises when, with Einstein in the van, some scientists deny the objectivity of free will on a similar ground. So much in the way of advance warning about what that meteor has suddenly brought back into the center of our cultural consciousness.

What about the origin of purely organic life? Does not the Bible, and its very first chapter indicate that God himself produced life as he went through the six-day creation? He did nothing of the sort. He rather said, as if endorsing evolution on the Earth (or on Mars!): "Let the earth bring forth vegetation," and indeed all kinds of vegetation.

Of course, Genesis 1 has nothing to do with evolution, nor with the special creation of each plant and animal species, for that matter. Genesis 1—and it is very difficult to explain all this during an interview with reporters interested not in serious thinking but in catchy phrases—is about the sabbath observance. A document from post-Exilic times, which saw the birth of orthodox Judaism with its visceral insistence on the sabbath, the author of Genesis 1 simply presents Almighty God in the role model of observing the sabbath.

Therefore it was most fitting to assign to God the greatest conceivable work, the making of the *all* or heaven and earth. The idea of *all* is repeated in the work of the 2nd and 3rd days, or the formation of the main parts (the sky and the ground) of the Hebrew universe, which is a huge tent. The work of the 4th and 5th days restate the *all* in terms of the main particulars of the upper and lower parts. Just think of the metaphor, "lock, stock, and barrel," to catch a most important point: a metaphor can literally state something, which is not metaphorical. Genesis 1 literally and repeatedly states that God created *all*, though it does not say that God created separately every thing. God, of course, does all this with supreme ease, as if with a flourish, which may very well be the meaning in that context of the verb *bara*.

The land animals should have come with the whales and the birds on the fifth day as the principal parts of the ground region, but they are introduced on the sixth day, so that the superiority of man as the manager of the world-tent, may have a proper backdrop. For if man has to be a worthy manager of God's greatest work, he has to have some special relation to that most intelligent of all architects, who is God. After all this, it may sound at least logical that such an architect should start his work by first producing light. Let the poor widow of the Gospel not be forgotten. She was intelligent enough to first strike a candle before looking for her lost coin.

One could only wish that countless exegetes of Genesis 1 had seen what should be obvious after so many false tries, recounted in my book, *Genesis 1 through the Ages*. In that case they should not have been bogged down in hairbrained stretching the physics of the day to get visible light independently of the sun.

In speaking in such a vein to reporters about Genesis 1, I meet with gasps of disbelief. They find it difficult to believe that I am sticking with the Bible. We, of course, must stick with the Bible, according to which man is created in the image of God, and therefore ought to be intelligent. Now, it was already recognized by Saint Augustine that God cannot reveal something to man that would be contrary to what man can clearly establish with his reason about his physical surroundings. This is why in patristic times the biblical flatness of the earth was quietly abandoned and the sphericity of the earth, as established by Greek science, was accepted.

Unfortunately there was not enough reflection on that inevitable reinterpretation of the Bible. Had that been done, the Galileo case might have been forestalled and all unnecessary sparring with Darwin and evolutionists avoided.

For there is nothing in the Bible or rather in the Faith that would impose on us any timidity to accept the really sound part of the science of biological evolution. That part is limited to the following proposition: the powers of matter are sufficient to explain all that is material in the rise of life and in any further differentiation of life into countless new species, vegetative and animal.

About all that is material in that process empirical science (a marvelous manifestation of the human intellect) is the first and last arbiter. I repeat material, that is observable and measurable. This is why it makes no sense on the part of a scientist to speak about purpose in life processes. Purpose is not an object of measurement or empirical verification. By the same token, it makes no scientific sense to deny purpose anywhere or everywhere in the universe. Purpose is a matter for philosophy, whether to assert it or deny it. This is even more so about intelligence. IQ measurements are irrelevant in this respect.

Everything which is measurable and empirically verifiable in evolutionary theory is the strict domain of science. And herein lies the rub for evolutionists. Evolution, as defined above in reference to the powers of matter, is much more a philosophical proposition than a scientific one. It demands the powers of the mind to assert connections where empirical evidence is sorely missing. The science of evolution is full of huge gaps, which today even people like Steven Gould openly admit. What they fail to see is that those gaps can be bridged only by a philosophy which it has been the stated purpose of materialist evolutionists from Darwin to Dawkins to discredit. But in the absence of that philosophy scientists as such can gain no better hold on the magnificent process of evolution leading from amoebas to man than one can grasp "the morning mist."

This expression "morning mist" was the only thing which one of the reporters, grilling me about the meteor, cared to put verbally in my mouth. Once more it became true that reporters are much more interested in catchy phrases than in sound reasoning, however elementary.

I have no doubt that this phrase of mine, "morning mist," will be used against me by those who do not care to look up the context. I can foresee the day when some science writers who deal with evolution and religion will recall that phrase of mine as typical of the dislike of "even members of the Pontifical Academy of Sciences" of evolution. So be it.

When about four years ago my remark, "I could not care less about the Big Bang," appeared in almost a hundred US newspapers, none of them referred to the immediate context. The broader context was the finding of strange discontinuities in the 2.7°K cosmic background radiation. The finding, which gave a new ground for theories about the formation of galaxies, filled a big gap in theories of cosmic evolution. Then the finding was taken for a proof of the creation of the universe.

Well, there can be no scientific proof of creation out of nothing, for the simple reason that the nothing cannot be observed and measured. It is in that sense that I could not care less about the Big Bang. And essentially for the same reason, I am not fazed by that meteor.

At any rate, if there are other civilizations elsewhere in our galaxy or in other galaxies, the intelligent beings there are to be accounted for in one of two possible ways. Either they are the chance product of darwinistically conceived evolution, or they are, as we are, so I believe, the product of God's direct creative action. In the latter case, their intelligence will be similar or analogous to ours and therefore communication between them and us will be possible even though they may not speak English. Moreover, they would also have free will and moral responsibility. Then they would not, without some pangs of conscience, take us for a convenient reservoir of slaves, let alone for a protein reservoir. In that case they might even teach us a thing or two about how to interpret and use science properly.

Within the Darwinist scenario, all those other species of intelligent beings will be part of a grim struggle of life, with no holds barred. In that case we had better hide. So much for the strange inconsistency in the enthusiasm with which some scientists and science reporters and some leading politicians (who try to make hay of everything in less than three months before election) greeted the news about ALH84001. That meteor tells a tale that has more than one exciting as well as frightening page to it.

Frightening also in a tragicomical sense. The latter was captured by a cartoon in *The Baltimore Sun*. It shows Dr. McKay, facing a group of reporters. He reads the following text, by itself a parody on the NASA scientists' overconfidence: "We have some *circumstantial* evidence that *might* point, in a *tentative* way, to the *slight* possibility that *maybe* 3 billion years ago there's a *chance* that there *could* have been-*ahem*-life on Mars." The other half of the cartoon shows a dozen reporters, who, on hearing this, burst into a frantic jubilation. Theirs is hardly the last laugh. But should they care? Newsmaking is based on the proverbial shortness of human memory.

10

Cosmology: An Empirical Science?

A title cast in the form of a question would suggest a negative answer. But even a mere hint that scientific cosmology is not an empirical science may be countered with the question: Should one doubt, however slightly, the empirical character of a science which ranges in time over the awesome stretch of 60 orders of magnitude and in space over 15 or so billion light years?

It may indeed seem foolish to set even a limit to the empirical character of scientific cosmology. Apart from dealing with the largest objects of which one has empirical evidence, scientific cosmology has also become indispensable for the study of the smallest bits of matter, usually called fundamental particles. One may indeed say that scientific cosmology is the most encompassing form of empirical

Invited paper for the Summer School on cosmology, Universidad Complutense, San Lorenzo di Escorial, July 1994. The Spanish version was published in J. A. Gonzalo (ed.), *Cosmología astrofísica: Cuestiones fronterizas* (Madrid: Alianza Editorial, 1995), pp. 248-270; The English text in *Philosophy in Science* 6 (1995), pp. 47-76. Reprinted with permission.

science and that all other empirical or physical sciences are, in part at least, mere subdivisions of scientific cosmology.

Yet there are some serious limits to the empirical character of scientific cosmology. Some of those limits are implied in some truly astonishing claims made by scientific cosmologists about their method and subject of investigation. Still another limit, inherent in the very peculiar nature of cosmology, is connected with the failure of scientific cosmologists to justify a very basic claim of theirs, a claim on which depends their very right to speak about a cosmos or Universe.

The most radical of those astonishing claims is that modern scientific cosmology deals with the creation of the universe, and indeed has the power to create. It would be difficult to assert any more strongly the limitless empirical sweep accorded thereby to scientific cosmology. No physical reality should seem more encompassing than the universe. Nor could one give a more exalted position to the empirical character of that science than by ascribing to it the power over that awesome act, creation, whereby the limit separating being and non-being, is crossed. It has for almost two thousand years now become customary in Western civilization to take the Latin word *creare* and its modern-language derivatives for an act whereby an entity is given existence in the strictest ontological sense. It is in that sense that the word "creation" has become synonymous with the idea which Christian theology, already around 200 A.D., wanted to specify by coining the phrase, creation out of nothing, or *creatio ex nihilo*. This idea, as can be readily gathered by consulting any good dictionary of any of the great Western languages, has become the primary connotation of the word creation. Intrinsically connected with that meaning is the assumption that only a being called God can create in such a way. In other words, insofar as creation out of nothing has for its object physical reality, it has been reserved to Almighty God's "empirical" power.

Against this linguistic and cultural (which is always cultic in one way or another) background it should seem rather surprising that not so much our poets (usually ready to take liberty with words) as our scientific cosmologists have laid a claim to that divinely empirical power. In fact so often and so many of them have done this as to "create" an "encyclopedic" truth. The *Encyclopedia of Astronomy and Astrophysics* contains, in its chapter on cosmology, three full columns[1] on what is the most limitless claim ever made within the context of an empirical science. The claim is that modern scientific cosmology possesses competence over the idea and fact of the creation of the universe. The claim is not the result of a momentary oversight about the proper use of words. It is thematically claimed there that the creation in question is a creation out of nothing. Now, if such a claim is reliable, it would be nonsensical to voice doubt, however muted, about the truly limitless sense of the empirical character of scientific cosmology. What can dispose more effectively of setting any bound or limit to empirical power than the ability to create and to create the *all* which is the Universe?

The word empirical has two basic meanings. The second of the two is markedly philosophical, or the proposition that knowledge is equal to the registering of sensory evidence. It is in that sense that one speaks of empiricism in philosophy. Scientifically more appropriate is the first of the two basic meanings, which in turn has two shades: In a broader sense empirical is that procedure which relies on experiment or observation. In a stricter sense empirical is a proposition which is capable of proof or verification by means of experiment or observation.[2]

There seems to be a general agreement that a science which has the ability to manipulate this or that kind or part of matter is empirical. If this is so, no empirical manipulation should seem to be more radical than the kind which does not merely transform matter, but brings it forth from nothing. Supremely empirical should seem that scientific cosmology about which it is now claimed that it can manipulate the totality of matter and indeed more than one such totality by creating a large number of universes out of nothing.

Few scientific cosmologists make that claim in more extravagant a manner than Professor Guth of MIT. He emphatically asserted to a reporter of *The New York Times* that the modern science of cosmology possesses, in theory at least, the empirical power to produce entire universes "literally out of nothing." And, hardly in jest, he added that our universe may have been produced in a basement laboratory in another universe.[3] It may, of course, be appropriate that this claim graced a Tuesday Science Section, as if even that illustrious daily could not do without some funny pages every week. But Professor Guth has claimed in scientifically respectable contexts as well, that the modern science of cosmology enables any able practitioner of it to create universes literally out of nothing.[4] Compared with this, the manner in which some molecular biologists claim to themselves the role of "playing God" may appear, from the empirical viewpoint, more immediately harmful but not nearly as sweeping.

The same sweeping claim is made, though in an untheatrical manner, in the *Encyclopedia of Astronomy and Astrophysics*. There one is told that "speculation qualifies as scientific if it is vulnerable to refutation by observations. The conjecture creation *ex nihilo* surely qualifies." The reason given is that it is possible to refute or verify whether "the universe has a net zero value for all conserved quantities. But since it is observed that matter [over radiation] is dominant, the baryon number cannot be looked upon as conserved." Therefore, so goes the conclusion, the conjecture of creation *ex nihilo* remains scientifically viable. Later, inflationary theory is praised because "inflation greatly enhances the plausibility of creation *ex nihilo*."[5]

Equally respectable, from the technical viewpoint, is the cosmological context where the supreme power assigned to modern scientific cosmology takes on both a picturesque as well as a thematically philosophical character. The scientific status of Professor Fang Li Zhi, who came into the focus of Western consciousness in connection with the Tiennenman Square incident, needs no recommendation. In view of our previous remarks about the history of the word "creation," no further warning is needed about the title, *Creation of the Universe*, of his book on scientific cosmology, which he co-authored with Li Shu Xhian. There the last chapter has for a title "Physics of the First Move." Those who find this phrase curiously reminiscent of the expression, "Prime Mover," and therefore of metaphysics, will not be surprised by what follows in that chapter. The two distinguished authors offer an "exit from the hell of Big Bang,"[6] once more playing metaphysicians and theologians, and indeed God, by parading in white lab-coats.

The hell in question is that act of creation which, through two thousand years of Christian theological tradition, formed the only logical basis of a heaven to come. Whether the Big Bang model should or should not be taken for a proof of that creation will be seen shortly. It is another matter that the Big Bang is evocative of a first moment. Here let it be noted that if heaven is taken for hell, a similarly arbitrary linguistic game will not fail to turn up in the next breath. Indeed, the two distinguished Chinese astronomers invoke Taoist philosophy, which is based on the obliteration of the difference between being and non-being. As a result, the universe is claimed to have come forth from nothing, with the proper assistance of modern scientific cosmology. And lest those who can read only pictures should miss the point, the two authors offer, perhaps because of their affinity with Chinese ideography, the picture of the flag of the universe. The flag carries the word NOTHING and nothing else.[7]

What a pity that an ideographic picture of the non-empirical content of "nothing" has not been provided! This, of course, should cause no surprise. It is hardly man's empirical powers that enable him to form the notion of something, a notion which is most emphatically non-empirical. Western man may, of course, be taken aback at being asked to give up age-old Western linguistic rules about the difference between nothing and something. Indeed, as we shall see, it is claimed that scientific empirical cosmology forces us to gloss over that difference and on the most encompassing scale at that.

Under the flag of the Universe, graced with the word NOTHING, a justification is offered on behalf of that claim in the form of a quotation from André Linde, another in the galaxy of modern scientific cosmologists: "The possibility that the universe was generated from nothing is very interesting and should be further studied. A most perplexing question relating to the singularity is this: what preceded the genesis of the universe? This question appears to be absolutely metaphysical, but our experience with metaphysics tells us that such metaphysical questions are sometimes given answers by physics." On looking up the context of this statement, made by Linde in 1982, one would search in vain for some specifics concerning that experience with metaphysics, or for the answers which physics has at one time or another given to metaphysical questions.[8] Of course, it is an old truth that whenever prominent physicists pontificate about metaphysics and theology they feel no need to give factual documentation or even show some modest familiarity with the record. Their empiricism extends only so far. It includes not a trace of doubt about the claim that scientific cosmology has gained supreme empirical power over everything that exists.

That kind of power should be suspect. Even the non-expert should be wary on hearing young physicists boldly claim that the modern science of cosmology, and quantum mechanics in particular, has abolished the difference between something and nothing.[9] When the chief fault found about a lecture on quantum cosmology is that the lecturer emphasized the difference between something and nothing, the doubts should not relate to the lecturer's logic. One should rather suspect the reasoning of the objector, or some pseudo-philosophical malaise administered to him in the name of empirical science.

The malaise has been raised to the level of encyclopedic truth in the *Encyclopedia of Astronomy and Astrophysics* where one reads that "quantum uncertainties suggest the instability of nothingness."[10] This, if true, means the reification of "nothingness" and, consequently, the abolition of the difference between nothing and something. Clearly, nothingness has to be something, if instability or any other property can be ascribed to it. If, however, physics is to progress in terms of a licence for the most arbitrary use of basic terms, one should at least be given the liberty to recall that such a licence was the hallmark of a land where physics (and cosmology in particular) had at a time been put in a straitjacket.[11] The land did not for too long survive the year, 1984, immortalized by Orwell. Well over a hundred years earlier, the arbitrary use of basic words received its poetical lampooning in Lewis Carroll's *Through the Looking Glass*. There Humpty Dumpty is free to declare: "When *I* use a word it means just what I choose it to mean,—neither more nor less."[12] In that land everything looks upside down and inside out, hardly the perspective from which to do good empirical science.

Indeed in an otherwise great scientific success there was something counter-empirical which marks the first step on the road toward empirical sleights-of-hand whereby entire universes, as if so many rabbits, are now being pulled out from under the physicist's linguistically cocked hat. The success was Heisenberg's formulation of the uncertainty principle. Like any great scientific achievement, this too looked almost like the unveiling of a great puzzle, although it was directly present in a basic characteristic of quantum mechanics. There one has to rely on non-commutative algebra as well as on the assumption that Planck's quantum is truly indivisible. From this it follows that there is a limit to the precision with which physicists can measure interactions involving parameters known as conjugate variables, such as position and momentum, energy and time.

Heisenberg, however, attributed a much broader significance to his justly celebrated formula. His famous paper came to a close with the words: "Because all experiments are subject to the laws of quantum mechanics . . . the invalidity of the law of causality becomes through quantum mechanics definitively established."[13] The word "definitively," that marks a stage usually not reached by a single bound, may by itself indicate that Heisenberg had beforehand serious doubts about the validity of causality. Indeed, Heisenberg and many of Heisenberg's colleagues had for some time rejected the principle of causality before his famous paper had been published.[14] For this they had to have grounds other than scientific. Since one observes only sequences, causality can be established or rejected only on philosophical grounds. Precisely because of this, if a philosophical meaning (in this case the overthrow of the principle of causality) is attributed to a formula in physics, the consequences will touch not so much on physics as on philosophy. More of this shortly.

In a less direct way, it was a combination of Kantianism and pragmatism that provided the philosophical reasoning on the basis of which Heisenberg felt entitled to make that momentous inference about the overthrow of causality by physics.[15] More directly, Heisenberg, like many other physicists, not only took measurements to be the mainstay of physics, but by developing an undue fondness for them, attributed to them an universal importance. Indeed, it was that undue fondness (understandable, though not forgivable in a physicist) that prompted Einstein to look for a greatly improved form of quantum mechanics that would allow measurement with full precision and restore thereby the principle of causality to its pedestal.[16]

Einstein, of course, tried to achieve the impossible. He failed to see that in Heisenberg's case the problem did not lie with physics but with philosophy. The same fallacious reasoning with which Heisenberg inferred the invalidity of the principle of causality, could not restore its validity by falling back on the reverse of the same fallacy. In Heisenberg's case the fallacy was the claim that as long as physicists could not measure interactions with perfect accuracy, the physical or material or ontological realm lacked causal coherence. In Einstein's case the fallacy consisted in the inference that if perfectly accurate measurements were possible, physical reality constituted a causally interconnected realm.

In more concise terms what Heisenberg thought to have established through his formula was not at all the incoherence of nature. He merely crowned with a scientific halo the following *non-sequitur* in logic: an interaction that cannot be measured exactly cannot take place exactly. The *non-sequitur* consists in the failure to notice that in the foregoing phrase the same word "exactly" is first used in an operational sense (cannot be measured exactly), and then in the sense that the interaction cannot take place, that is, exist, "exactly." In this second case the meaning of "exactly" is not operational but ontological. To go without further ado from the operational to the ontological is, however, the kind of misstep in elementary logic for which the Greeks of old, so keen on logic, had already a technical expression, *metabasis eis allo genos*, or the jumping from one realm to another, very different realm.

Heisenberg's paper provided a scientific seal on a wholly unwarranted anti-ontological inference. The pervasive use of that inference can best be seen in the writings of Niels Bohr. Even more than Heisenberg himself, Bohr served as the authoritative spokesman on behalf of what became known as the Copenhagen interpretation of quantum mechanics. It is enough to recall here a major study of Bohr's writings where attention is drawn to Bohr's systematic evasion or slighting of all questions about ontology.[17] It shows the philosophical poverty of the scientific establishment that the Copenhagen interpretation was hardly ever opposed with reference to ontology. In one notorious case (involving Born, Pauli, and Einstein) concern for ontology was brushed aside as no more meritorious than efforts aimed at specifying the number of angels that can be put on a pinhead.[18]

Such a cavalier handling of ontology and of the true coherence it alone can secure for things physical readily led to a celebration of incoherence. With a scientific coating given to that cavalier attitude, the result gradually established itself as an encyclopedic truth. A telling example of this is in a work of far greater academic distinction than its facetious title *Lying Truths* may suggest. Published in 1979, it deals with what is not known and what cannot be known in various fields of inquiry.[19] One of the contributions is by the cosmologist, P. C. W. Davies. The very title of his essay, "Reality exists outside us?" should make it of interest to philosophers and irrelevant to empirical science. It is, however, in the name of science that Davies banishes coherence from the physical realm: "The uncertainty of Heisenberg is an inherent *property of nature*." No statement could be more philosophical and less scientific, and certainly in the empirical sense. To heighten the irony, the statement is from a book unofficially referred to as the *Encyclopedia of Ignorance*. Something elementary has indeed been ignored by Professor Davies and countless colleagues of his. They were ready to ignore also some monumental fallacies soon to envelop the cosmos itself.

The beginnings could not have been more minuscule. Only a year after Heisenberg enunciated the uncertainty principle, George Gamow made a memorable use of it in explaining the emission of alpha rays from the nucleus. Gamow was concerned with the fact that alpha particles can escape from the radium nucleus although their kinetic energy is small compared with the energy barrier surrounding that nucleus. Had Gamow been concerned with the time of emission, he might have perceived the real point. If he failed, it was partly because of his notorious scoffing at philosophical considerations. He had no sensitivity to the fact that the ontological distance between being and non-being is infinite, even if the entity in question is a mere fraction of the mass of the alpha particle. But unless there is sensitivity to that non-quantitative infinite distance, there will be no adequate concern for the equally infinite distance between being and boing when it involves the total mass of the universe.[20]

Such a fraction of mass, perhaps of the order of 10^{-34} gm, is on hand when one tries to measure the moment or time of the emission of an alpha particle, say, from a radium atom. The measurement is subject to the uncertainty principle which can be written not only as $\Delta x \cdot \Delta mv \geq \hbar$ (where x is position, mv is momentum and $\hbar$ is Planck's quantum divided by 2π), but also as $\Delta E \cdot \Delta t \geq \hbar$ (where E is energy, t is time). But since E can also be written as mc^2, ΔE becomes Δm (or an error in measuring the mass of the alpha particle), because c or the speed of light is invariable. In itself all this means only that there is a limit to the precision of measuring Δm. However, something entirely different is meant if one grants the Copenhagen interpretation which implies the inference that what cannot be measured exactly, cannot take place exactly. Then Δm, or the error in measuring m, becomes a real or ontological mass defect. But with a studied neglect for ontology (or metaphysics) on hand, neglect of something all important in physics began to countenanced. Indeed, a brazenly new leaf was turned in the book physics: the need to balance the sheets was no longer considered paramount.

Within less than two decades this lackadaisical stance took on the air of unquestioned respectability among cosmologists. They took it in stride, indeed many of them rejoiced, when in the late 1940s the steady-state theorists came up with the claim that every second one hydrogen atom pops up into existence within a space as large as a cubic mile. They did not press on with the question about the source of that puny bit of matter. Perhaps they felt that there was no need to balance the books of physics if the imbalance, with respect to the principle of the conservation of matter, was merely 10^{-28} gr of matter, per cubic mile, per second? Of course, if one had summed up the matter thus gained for nothing throughout the then known realm of galaxies at every second, the sum total would have amounted to several sun-masses, hardly a negligible quantity. The original proponents of the steady-state theory made no secret of their view that all those hydrogen atoms could be obtained for nothing. They popped up, so the claim went, not through some unknown physical process or from some recondite form of matter, and certainly "not out of radiation but out of nothing." In making this specification, Bondi certainly did not mean to invoke a Creator![21]

The claim is, of course, a rape of logic. Compared with that claim the theological doctrine of creation should stand out, as what it truly is, a mental operation in which logic is given its full due in two respects. First, according to sound theology, the act of creation out of nothing is anything but observable, whatever the full observability of matter that has already been brought to existence. No less importantly, since such an act touches at the very root of existence, it can be performed only by the One known as HE WHO IS, that is Existence itself. Indeed, God is so unique in that respect that, according to theology, even his omnipotence does not enable Him to communicate that creative power of His to any creature, however exalted. A crash course in theology may stand in good stead some cockeyed members of the cosmological confraternity, specializing in exalted sleights-of-hand.

The farcical nature of that specialty can, of course, be stated in a language, which, relating as it does to the world of banking, may be more understandable than the language of theology. For the astonishing claim that modern scientific cosmology is competent to create a universe and indeed delivers the power to create as many universes as desirable, is equivalent to the setting up of a most gigantically dubious World Bank. This bank is not merely global but one where transactions have the Universe and even more than one Universe for their object. By a much too transparent manipulation of Heisenberg's principle modern scientific cosmologists claim that from this World Bank they can draw matter equivalent to the mass of the entire universe without first securing the deposit of that amount.

Perhaps in this age of universal and computerized banking this analogy may convey a pseudoscientific farce to those who can no longer be budged by philosophical warnings, however appropriate.

In view of all this, something redeeming may be seen in a memorable disclaimer made by Gamow in 1952. It came on the heels of the great success of his *Creation of the Universe*, first published a year earlier. There he gave to the wider public an entertaining glimpse of the physics in a very hot and very condensed early universe. Contrary to his fully agnostic intentions many took his book for a scientific proof of creation. He therefore introduced the second edition with a most sensible warning: "In view of the objections raised by some reviewers concerning the use the word 'creation,' it should be explained that the author understands this term, not in the sense of 'making something out of nothing,' but rather as 'making something shapely out of shapelessness,' as, for example, in the phrase 'the latest creation of Parisian fashion'."[22] One could only wish that Gamow had added that neither fashion designers, nor physicists can ever start from anything truly shapeless. Are not gluons and quarks—with their charms, colors, and flavors—exceedingly shapely entities?

Unfortunately, Gamow's wise disclaimer was not heeded either by the wider public, or, in a different sense, by two subsequent generations of scientific cosmologists. Not a few them grew fond of claiming that they can create in the very sense in which Almighty God alone can. Whatever the theological blasphemy on hand, science is blasphemed when a patently fallacious claim is called empirical science.

Slightly less outrageous, though still very astonishing, is the claim that science can prove the eternity of the universe. The claim is empirical only in the sense that whenever a scientist or a group of scientists makes a broader claim, not at all empirical, it quickly takes on an empirical aura. This happened after the steady-state theorists came forth with their perfect cosmological principle, namely, that a universe that appears homogeneous in space, should also look homogeneous across time. The principle was aimed at offsetting the effectiveness whereby the empirical evidence about a universe in expansion evoked the specter of its temporal beginning, and beyond that its very creation. Fred Hoyle certainly made no secret of the anti-theological motivation which prompted him to espouse that principle. He certainly succeeded in selling the idea that an empirical observation of a surplus of 21cm radiation (the natural radiation of hydrogen) in cosmic spaces would prove the popping up of hydrogen atoms out of nothing.

Nobody in the scientific community was enough of a philosopher to note that no such surplus radiation, and indeed no observable physical process can ever be taken for an empirical proof of a transition from non-existence into existence. Such a transition, or creation out of nothing, remains strictly unobservable. That the universe owes its ultimate origin to a creation out of nothing is a philosophical inference based on its metaphysical contingency and not a conclusion from empirical observation.

Today, little needs to be said about the eternity of matter as a basic "scientific" dogma within the once powerful political establishment of dialectical materialism and within its Western circle of admirers, many of them in prominent scientific positions. But one case may be recalled about the grim resolve with which even well-meaning scientific cosmologists had to pay their due to that scientistic dogmatism, by claiming empirical status to it. Professor Ambartsumian arrived at the last moment to the 17th World Congress of Philosophy, held in Düsseldorf in August 1978. He came to deliver an invited paper as a member of the cosmology panel. As he read his heavily technical paper on star formation, all of a sudden he declared that the most reliable conclusion empirical science had ever established is the eternity of matter. Obviously he wanted to satisfy two husky compatriots of his, officers of the Soviet Secret Police (KGB), who accompanied him and took their seats in the front row as corresponding members of the Uzhbekistan Academy of Sciences. To satisfy them and the Party, Ambartsumian had to pay homage to the Marxist dogma of the eternity of matter (universe), the basic dogma of all materialists.

As a member of the panel, I had the opportunity to make a comment before questions were taken from the floor. I asked Professor Ambartsumian whether the experiment that would provide the proof of the eternity of matter should not itself be of eternal duration. He was visibly taken aback when I added that I would wait for his reply until hell freezes over. I am still waiting.

Equally unredeemable is the claim, very fashionable among scientific cosmologists, who do not want to be taken for rabidly dogmatic materialists, that a temporal beginning for the universe is less aesthetic than its existence without a beginning. Aesthetics is one thing, empirical evidence is another. Unfortunately, here too, the espousal by scientific cosmologists of a patently non-empirical postulate can easily take on an empirically respectable status.

This happens even more readily when the pseudo-metaphysical longing for an eternal universe is couched in a mathematical formula. The proof is the avidity with which an idea of E. P. Tyron, proposed in 1978, has been seized upon. It was certainly an interesting finding that the rest-mass energy of the total mass in the universe and the gravitational attraction of that mass are nearly equal. But since the former is a positive and the latter is a negative quantity, their sum is close to zero. Feeding this result into Heisenberg's uncertainty principle, written as $\Delta t \geq \hbar/\Delta E,$ one obtains on the right side of the inequality a fraction where a small quantity (Planck's quantum) is divided by an even smaller quantity. The result is a very large number, though anything but infinite. In spite of this, it has become the fashion, since 1978, among scientific cosmologists to see in Tyron's idea an empirical proof of the eternity of the universe. Such was a patent mishandling of plain quantities. Doing it with impunity provides the assurance that one has therefore on hand not only eternity by formula but also an empirical proof that the universe is uncreated.

Totally ignored in the process has been the fact that the universe thus obtained cannot be more than a virtual universe, similar to other virtual particles that can exist only within the time limit set by Heisenberg's formula. But even that existence assumes the existence of at least two real universes between which the virtual universe can act as an exchange particle. In other words, even in good physics, to say nothing of good philosophy of knowledge, the first step is truly real matter and not merely virtual matter or universe. Furthermore, if one wants to have, by virtue of that formula, an eternally existing virtual universe, one must first have on hand at least two eternally existing real universes. Or else one would be flouting all physics about virtual particles. But then where is the empirical proof of an eternal universe? Or should one take the art of putting the cart before the horse for empirically good scientific cosmology?

The self-defeatingly furtive presence of two real universes in that "proof by formula" of the eternity of a universe only virtually existing is not something to be taken lightly by anyone respectful of logic. Otherwise one will overlook the fact that logic is handled rudely in still another astonishing claim made on behalf of the empirical sweep of modern scientific cosmology. The claim is that it deals with not one but with several and, indeed, with uncountably many universes.

Logic, to say nothing of science, is honored in the breach in the so-called multiworld theory of the universe. In itself, the multiworld theory does not state the multiplicity of universes. The reason for this is that the multiplicity of universes has a touch of realism to it which the multiworld theory cannot include. Clearly, if physical reality is made dependent on individual consciousness that makes the wave function "collapse," no convincing case can be made for realism. Apart from this, the multiworld theory fails to explain why so many individual states of consciousness can effect the realization of one single universe. Much less can one find an explanation in the multiworld theory for the existence of physical reality prior to the emergence of conscious beings.

Should one say that the primitive consciousness of dinosaurs and, before that, a proto-consciousness of amoebas, and, prior to all that, a primordial consciousness of hydrogen atoms, or of quarks and gluons, achieved the collapse of their uncountable wave-functions into reality? Or should one assume that the thinking of physicists in other universes about the wave-function of all particles in our universe made it "collapse" into reality? But then there still remains the foregoing series of questions about the consciousness of dinosaurs, amoebas, gluons and quarks in other universes. Or should one recast physics in terms of Hegelianism, where the World-Spirit comes first and, as Hegel's *Enzyclopedie der Wissenschaften* shows,[23] makes all good physics impossible?

That such and similar questions do not figure in the wide literature on the subject does not reflect well on the philosophical acumen of scientific cosmologists. They have shown no concern either for a problem less exacting epistemologically, a problem relating to the proper use of the term "universe." Here again, let the apparently scientific aspect of the problem be dealt with first. It is not science, let alone empirical science, to turn, as is done in the inflationary theory, our universe into one universe within an ensemble of universes, all generated statistically and even with different sets of laws and constants. Science begins where quantitative forms are given to propositions. What about universes where the speed of light is not constant, or has a value different from the one observed in our universe? What about universes with no gravitation? What about universes composed of matter wholly unimaginable to us? Can scientific, that is, empirically or quantitatively verifiable answers be given to such questions?

Apart from this, it is a mere evasion of a problem to postpone endlessly the task of coping with it. The problem of other or multiple universes is not solved by taking refuge in a super-ensemble of universes. Is that super-ensemble the true Universe, or merely a part of another even larger agglomerate of universes? Is it empirically respectable to postpone the answer forevere? Is not scientific cosmology vitiated by the expediency of infinite regress? Does not scientific cosmology too need a desk, such as the one in the White House, about which a very down-to-earth President, Harry Truman, spoke as the place where the buck stops?

Mere respect for basic definitions should make one wary of brave references to "other universes." Does the word universe stand for the totality of things, or for something else? This problem is merely evaded, and very cheaply at that, by falling back on the distinction between the strong and weak sense of a word, a technique which all too often mars scientific discourse. At any rate, if two or several universes are in a scientifically meaningful interaction with one another, they obviously form only one universe. (If they are not, they are unknowable to one another). This is a point still to be widely recognized in reference to other universes that are said to float across the cosmological horizon into our universe in some models of relativistic cosmologies. The very reference to floating should evoke some elementary truths about hydrodynamics. Two rivers can flow into one another only if both are ruled by the same laws of physics. The same cosmic hydrodynamics must also be common to universes that are supposed to float across each other's horizon. But if they have so much in common as to be able to join up with one another, why should they be looked upon as other universes?

The foregoing discussion of some astonishing claims in modern scientific cosmology should reveal some very stark limits to its empirical competence. That competence cannot have for its object the creation of the universe out of nothing, or the determination of its eternity, or of its multiplicity.

To what extent can scientific cosmologists justify a far less, though very scientific claim of theirs, namely, that they deal with the quantitative properties of a true totality of things, or Universe, already existing? The writings of modern scientific cosmologists reveal a curious reluctance to face up to the problem of whether the true totality of matter, or the Universe writ large, is the object of their investigations. Yet a serious consideration of this problem could have cast light on the sense in which scientific cosmology is a particularly non-empirical science whatever the enormous range of its empirical powers.

A scientific cosmologist cannot claim to have pondered at any length the question, "How do you know that there is a universe, a strict totality of things?" if he or she blithely answers: "It is self-evident that there is a universe." It is perhaps because they take the Universe for something self-evident that scientific cosmologists do not care to consider the question at all. Thus in *Origins*, a collection of interviews with leading modern astronomers, only once did the question come up of how one does know that there is a Universe. The reply, given by A. Lindé, was ambiguous: "Different people use this word differently." He himself thought that consciousness should be included in the definition of the physical universe.[24] Is not this an unintended invitation to land in the murky waters of the multi-world theory?

The word "self-evident" is best left aside, as it may turn out to be a philosophical landmine, ready to explode. Let us focus on the expression "empirically evident". Even there some philosophical caution may be in order. Once I heard a world-famous astronomer state that the stars we see through our telescopes exist only on our retinas. To my question, whether he would say that the wall in front of him exists only on his retina, he said that the reply demanded some reflection on his part.[25] Indeed, more than a little reflection is needed to gain rational assurance that there is a Universe.

Such an assurance demands much more than, say, to ascertain empirically the existence of a spherical earth. This can be done by circumnavigating it. A fully direct evidence about the earth's sphericity was obtained when astronauts orbited around it. Empirically less direct, but still heavily empirical is our conviction about the existence of the solar system, of our galaxy, of systems of galaxies, and supersystems of galaxies.

But what is our empirical evidence about the existence of a universe taken for the strict totality of all material entities? It would be presumptuous to take the confines of the visible universe for that totality. This would turn the universe into a function of the gradual improvement of telescopes. At any rate, well before one reaches the realm of a billion light years, estimates of distances become fraught with uncertainties. Even today, after spectacular successes in improving distance estimates, a now over half-century old remark of Edwin Hubble's remains worth pondering: "With increasing distance, our knowledge fades, and fades rapidly. Eventually we reach the dim boundary—the utmost limits of telescopes. There we measure shadows, and we search among ghostly errors of measurement for landmarks that are scarcely more substantial."[26]

Hubble may have sounded too skeptical, but his warning must be kept in mind when it comes to evaluating, on the basis of relativistic cosmology, the total mass of the universe. The empirical basis of that computation is the estimate of the average density of matter. This in turn depends on the counting of galaxies in a given volume of space. This estimate also allows one to compute the path of permissible motion with the minimum curvature. But is the radius of that curvature the radius of the Universe? There is no empirical reason why these data should be looked upon as being in a one-to-one correspondence with the Universe as such. In other words, is Einsteinian cosmology, the very basis of all scientific cosmology, really about the Universe writ large?

Yet, it has been tempting to answer this question in the affirmative. Over vast stretches of space and time the science of cosmology sees an immense amount of matter obeying the same laws, the same overall expanding motion. This unitary view is further enhanced by the fact that scientific cosmology can follow the evolution of that colossal amount of matter from exceedingly small and very early states to its present state. This success is all the more remarkable and all the more expressive of a cosmic unity as it implies, at its very early phases, the stepwise unfolding, or freezing out, of four different forms of forces as if they were but different aspects of one and the same force.

Since the COBE (Cosmic Background Explorer) experiment, scientific cosmology has filled, so it seems, the last great missing piece in its picture of the evolution of the cosmic whole. Slight variations in the cosmic background radiation now give theoreticians an empirical grip on the formation of the first galaxies. Perhaps only minor details remain to be worked out in an all encompassing picture. Still, the picture, however complete, will not contain an intrinsic proof that it is *the* picture, the empirically guaranteed picture, of the Universe, writ large.

The proof cannot be given empirically. Unlike other branches of physical or empirical sciences, scientific cosmology cannot have an overall empirical knowledge of its object, which is the Universe writ large. While scientific cosmology can account for trillions of trillions of galaxies, on no account can it take their totality for the Universe as such. No scientific cosmologist can ever devise an experiment that would enable him to observe the Universe from the outside. It is just as impossible to get outside the universe as to get outside one's very skin.

Is there a theoretical way of knowing that there is a physical Universe that contains all physical entities? Can a so-called final theory, which so many leading physicists from Einstein on tried to formulate, provide the answer? Such a theory must at least contain something about the universe, which is, however, not the case with Steven Weinberg's book, *Dreams of a Final Theory*. Fifty years after Gödel had formulated his incompleteness theorems, Professor Weinberg still does not seem to know them, or at least something about their relevance to final physical theories. But as long as those theories are valid, no scientific cosmology can claim consistency to itself.[27]

Should we therefore speak of galactology instead of cosmology? As I said above, the unitary and encompassing picture which modern scientific cosmology unveils about the world at large, must not be taken lightly. It is a picture very evocative of that coherent whole which is the universe. But still that picture, however grandiose and consistent, cannot be taken, without some reservation, for an account of the totality of things across space and time. Keeping this difference in mind would prevent scientific cosmologists from engaging in a dubious form of metaphysics in the guise of doing superb physics. Such a most dubious form of metaphysics is being construed by scientifically coated talks about the eternity of matter and about a scientifically performed creation of matter out of nothing.

Yet one need not be immersed in arcane and very abstract forms of metaphysics in order to gain rational assurance about the empirical reality of a totality of things or the Universe. Mere assertions, based on some form of process philosophy, will not suffice and indeed serve as so many red-herrings. Popper, Alexander and Whitehead (to mention only some notables) never proved that there is a Universe, yet they asserted without hesitation that the Universe takes up all forms of existence as it unfolds itself through eternity.[28] If one has not yet proved that there are animals, it makes no sense to speak about their evolution.

The very closeness of two words, universals and universe, may suggest that a reliable philosophical proof of the existence of a strict totality of things may hinge on one's appreciation of the basic function of universals in all human knowledge.[29] But this is a paper on scientific cosmology. So let philosophy be set aside as much as possible. Let the answer be sought from the direction of empirical science, which stands or falls with its ability to count bits of matter. For every act of measuring of no matter how small or large an object is an exercise in counting. Further, empirical science is such because the empirical realm, or matter, is countable. Any bit of matter which is not countable ceases to exist for the purposes of science, which also claims to be empirical.

On the cognitive road to the Universe the starting point is the registering of the reality of matter which has countability as one of its chief properties. Counting in turn is done in terms of integers. Reflection on the nature of integers is the basis for the inference about the impossibility of an actually realized infinite quantity. It is in terms of that impossibility that one can assert the finiteness of a universe made of countable matter.[30] Scientific cosmologists, who speak volubly about the ability of hyperbolic space to contain an infinite amount of matter without giving rise to the gravitational paradox, seem to forget two things. One is that in four-dimensional manifolds other than spherical there can be no physically meaningful central field of force, the very reason for the gravitational paradox already in a three-dimensional world.[31] The other is that even if a hyperbolic manifold could accommodate an infinite amount of matter without incurring the gravitational paradox, the contradictory character of infinite quantity still remains. At any rate, to worry about infinite energy and to countenance at the same time an infinity of matter should seem a strange inconsistency to anyone mindful of the mass-energy equivalence.

It is the universal presence of integers as a basic condition for the countability of matter which guarantees an all-pervasive rationality for the purposes of scientific cosmology, or of any branch of physical science. It is the countability of matter in terms of integers that assures it the kind of coherence which science must assume about the material universe. It is up to empirico-theoretical investigations to find out the particular mathematical form embedded in the structure and interactions of matter. Therein lies the true objective of science and also the independence of its method from metaphysics.

The reverse side of that independence is the fundamental indebtedness of scientific cosmology to metaphysics for its cognitive starting point and for its supreme object, the cosmos. Metaphysical reasoning alone can assure the scientific cosmologist that there is a Universe, a coherent totality of consistently interacting material things. The Universe is ultimately not a question of empirical evidence but of metaphysical inference. Once this is admitted, no objection can be raised to the additional inference, whose object is the One to whom alone one can ascribe the origin of the Universe as a result of a creation out of nothing, a process which is anything but empirical.

[1] Robert A. Meyers, (ed.), *Encyclopedia of Astronomy and Astrophysics* (San Diego: Academic Press, 1991). pp. 155-57.

[2] See *Webster's II* (Boston: Houghton Mifflin Co., 1984), p. 428.

[3] In an interview with M. W. Browne, "Physicist Aims to Create a Universe, Literally," *The New York Times*, April 14, 1987, pp. C1 and C4.

[4] For instance, in the article, co-authored with P. J. Steinhardt, "The Inflationary Universe," *Scientific American*, April 1984, pp. 116-28, where the word "absolutely" further strengthens the otherwise strong word "literally."

[5] *Encyclopedia of Astronomy and Astrophysics,* pp. 155-56.

[6] Fang Li Zhi and Li Shu Xian, *Creation of the Universe*, tr. T. Kiang (Singapore: World Scientific, 1989), pp. 145-47.

[7] Ibid., p. 147.

[8] The statement is from Linde's paper, "The New Inflationary Universe Scenario," published in *The Very Early Universe: Proceedings of the Nuffield Workshop, Cambridge 21 June - 9 July, 1982* (New York: Cambridge University Press, 1983), pp. 205-49.

[9] A graduate student, following my lecture on cosmology and creation, given in April 1988, at California Institute of Technology.

[10] *Encyclopedia of Astronomy and Astrophysics*, p. 157.

[11] For details, involving the execution of prominent cosmologists, see my book, *The Relevance of Physics* (Chicago: University of Chicago Press, 1966; paperback reprint: Edinburgh: Scottish Academic Press, 1992), pp. 487-88.

[12] L. Carroll, *Alice's Adventures in Wonderland* and *Through the Looking-Glass* (New York: New American Library, 1960), p. 186.

[13] In German, "Weil alle Experimente den Gesetzen der Quantenmechanik ... unterworfen sind, so wird durch die Quantenmechanik die Ungültigkeit des

Kausalgesetzes definitiv festgestellt." See, W. Heisenberg, "Ueber den anschaulichen Inhalt der quantentheoretischen Kinematik und Mechanik," *Zeitschrift für Physik* 43 (1927), p. 197.

[14] See P. Forman, "Weimar Culture, Causality and Quantum Theory, 1918-1927," *Historical Studies in Physical Science* 3 (1971), pp. 1-115.

[15] As abundantly clear either from Forman's essay or from Heisenberg's books, *Physics and Philosophy* and *Physics and Beyond*.

[16] This was the gist of a dispute, in the 1930s, between Einstein and Bohr concerning the question whether it was possible to devise a thought-experiment that would allow perfectly accurate measurements.

[17] As convincingly shown in C. A. Hooker's essay, "The Nature of Quantum Mechanical Reality: Einstein versus Bohr," in R. C. Colodny, (ed.), *Paradigms and Paradoxes: The Philosophical Challenge of the Quantum Domain* (Pittsburgh: University of Pittsburgh Press, 1972), pp. 67-302.

[18] The remark is Pauli's who thus characterized Einstein's concern for realism. For details and documentation, see my *God and the Cosmologists* (Edinburgh: Scottish Academic Press, 1990), pp. 123-24.

[19] The essay is part of *Lying Truths: A Critical Survey of Current Beliefs and Conventions,* ed. R. Duncan and M. Weston-Smith (Oxford: Pergamon Press, 1979), p. 149.

[20] That this point can be wholly lost on the modern physicist or cosmologist is best illustrated by a claim of D. Sciama (quoted in *Origins,* p. 142; see note 24 below) that "continuous creation of matter is even less a thing to introduce than the creation of a whole universe at one go."

[21] See. H. Bondi, *Cosmology* (2d ed.: Cambridge: University Press, 1959), p. 144.

[22] G. Gamow, *Creation of the Universe* (New York: Viking Press, 1952), p. [vii].

[23] For some stupefying details from that work, see my Gifford Lectures, *The Road of Science and the Ways to God* (Chicago: University of Chicago Press, 1978), pp. 139-140. If Hegel's philosophy of knowledge is the pudding, then its disproof, if needed, is certainly in his account of the physical sciences.

[24] A. Lightman and R. Brawer, *Origins: The Lives and Worlds of Modern Cosmologists* (Cambridge, Mass.: Harvard University Press, 1989), p. 393.

[25] The astronomer was the late W. McCrea, the context the question-answer period following a lecture of his at a Conference at the University of Denver in November, 1984.

[26] E. Hubble, *The Realm of the Nebulae* (New Haven: Yale University Press, 1936), p. 202.

[27] I have made this point in numerous contexts, since I first did so in ch. 3 of my book, *The Relevance of Physics*.

[28] See ch. 6, "Hollow Metaphors," of my book, *The Purpose of It All* (Edinburgh: Scottish Academic Press, 1990).

[29] For details, see my book, *Is There a Universe?* (Liverpool: Liverpool University Press: New York: Wethersfield Institute, 1993), the expanded text of my Forwood Lectures given at the University of Liverpool in November 1992.

[30] The impossibility of an actually realized infinite quantity is to be distinguished from going on endlessly with counting, or with the possibility of dividing an entity into infinite parts. It is also not to be confused with Cantor's transfinite number which, to recall a remark of Hilbert, has nothing to do with the physically real world.

[31] For further details, see my *Paradox of Olbers' Paradox* (New York: Herder & Herder, 1969).

11

To Awaken from a Dream,
Finally!

Advanced reviews of Weinberg's book* were uniformly raving and so were all the reviews in periodicals that determine public perception. Yet the book was not half a year off the press when serious doubts arose about the fate of its author's immediate dream, the SSC or Superconducting Super Collider. A few months later, in mid-October 1993, the US House of Representatives aborted the SSC by a vote that nobody, in his waking hours, expects to be reversed, not at least before a decade or so.

Down the drain went 4 billion dollars already spent on the project, a very disappointing way of preventing another 10 billion or so from being disbursed during the next four or five years. Compared with those hefty sums, relatively puny could seem the hundred or so million dollars needed each year to run the experiment: two opposing beams of protons to be hurled at each other at 20 trillion electron

---

*Essay review of Steven Weinberg, *Dreams of a Final Theory* (New York: Pantheon Books, 1992). First published in *Philosophy in Science* 6 (1994), pp. 159-74. Reprinted with permission.

volts so that traces of Higgs bosons, tokens of presumed steps toward a Final Theory, might turn up.

A goodly part of those 4 billion dollars paid for the boring of a 12-mile stretch, or about a fourth of a quasi-circular tunnel in the Austin Chalk, an eighty-million-year-old sedimentary rock lying under Ellis County, Texas. Instead of high-energy physicists, that stretch will host prairie dogs, gophers, rabbits, and rats. They will race up and down without ever suspecting that protons were meant to whirl around there at almost the speed of light. Running around in a circle is now being done by prominent physicists who do not want to take much of the blame for the fact that the greatest experiment in the history of physics did not become a reality.

Not a few within the scientific community, who have priorities other than to bag esoteric particles, cheered the Honorable gentlemen on Capitol Hill who killed the project. Fundamental particle physicists must, however, have sympathized with Leon Lederman, a Nobel-laureate and former director of the Batavia accelerator (Fermi Lab). As he took on the congressmen who had killed the project, he voiced his astonishment that "a large number of fairly intelligent people could make such a dumb thing."

This was on Lederman's part a not so subtle exercise in the art of damning with faint praise, and an all too transparent example of going on the attack as the best form of defense. The tactic was certainly needed if the Batavia group directed by him had been really guilty of not letting a plethora of physicists, needed by a new major project, encroach on their very turf. The Batavia group may now feel ill-concealed satisfaction over the fact that their leader, Dr. John Peoples Jr., was appointed to supervise the closing down of the SSC at a tune of at least a billion dollars. Yet all those at Batavia must have known that to expand an already existing accelerator would have cost much less than to build a much bigger one from scratch. As to other and lesser expenses, such as 50,000 dollars spent on potted plants, to say nothing of the tab of lavish parties, they too are common misdemeanors in big science.

A sound sense of fiscal responsibility cannot be strong in a profession full of wizards with warped thinking. They feel they are entitled to a very comfortable lifestyle for doing the very specialized kind of theoretical work they have always wanted to do. The work is indeed so specialized that they are not reluctant to speak of it as a game, different as it can be from any spectator sport, though also highly paid. The game may be most interesting to them but admittedly of no interest and use whatever to countless others. This is not to suggest that those others are necessarily enlightened in their efforts to restrict public research money to "socially meaningful" projects. In the face of one such effort I squarely endorsed continued work on the SSC, following a lecture of mine on cosmology and final theory in New York, in February 1993. However, I added that the project was not necessarily a step toward the final theory because the latter can never be found with convincing certainty.

Surprisingly, nothing is hinted about this last and most fundamental question concerning the SSC in Weinberg's book, which is essentially a public relations stunt, though on a very high level of science popularization. In that latter respect *Dreams of a Final Theory* surpasses Weinberg's earlier work, *The First Three Minutes*. But in a far more important respect Weinberg failed in both works. In *The First Three Minutes* he took cover under his scientific aura to make a grand but clearly non-scientific conclusion, namely, that there seems to be no purpose in the cosmos. That he expressed regrets on that score in *Dreams of a Final Theory* exposes there even more his failure to see something of the stringent limits of the scientific method even with respect to science itself. Equally silent on this point were all those rave reviews.

Yet the need to come to grips with those limits looms large almost from the very start. Weinberg has barely completed half a dozen pages when he claims that whatever the shortcomings of "present theories . . . behind them now and then we catch glimpses of a final theory, one that would be of unlimited validity and entirely satisfying in its completeness and consistency" (p. 6). That this very last word spells disaster for his dreams, he fails to notice. He is insensitive even to some obvious inconsistencies plaguing his diction. Clearly, if it is true that "we are not likely to discover it [final theory] soon," it cannot also be true that "from time to time we catch hints that it is not so very far off" (p. 6). In the same first chapter, called Prologue, Weinberg makes two more statements about the characteristics of a final theory. The final theory is "so rigid that it cannot be warped into some slightly different theory without introducing logical absurdities like infinite energies" (p. 17). The final theory would bring to an end not all science (or physics in particular) but only "a certain sort of science, the ancient search for those principles that cannot be explained in terms of deeper principles" (p. 18). In

Weinberg's own words, a wide range of "wonderful phenomena, from turbulence to thought, will still need explanation whatever final theory is discovered" (ibid). Such a concession is an unabashed bow to the ideal of reductionism which Weinberg later upholds through a whole chapter.

But before giving his "two cheers for reductionism" Weinberg portrays the confluence of two distinct types of considerations into the making of an ever broader and, eventually, final physical theory. Universal considerations, as he calls them, relate to the universal validity of a generalized statement or law. He calls historical those considerations that are more commonly called physical constants and initial conditions. Perhaps he should have called them the specifics of physical theory that most starkly reflect the enormous measure of the specificity of the physical universe. Tellingly, he does not like to speak of specificities. No wonder. Elementary logic would tell him that no specificity can be derived from the kind of universality which borders on strict homogeneity or perfect symmetry. A little more than logic, elementary metaphysics, would then tell him that precisely because the universe looks so specific, anyone with unlimited respect for sufficient reason would be entitled to raise the question, and about that totality which is the universe, why such and not something else.

This is the question which lands one at the doorstep of theology or religion. In the next to last chapter of the book Weinberg comes clean on religion. The only kind he would consider is panentheism, wherein the universe and deity are no more convincingly indistinguishable than in plain pantheism. No place in either for worship properly so-called.

At any rate, Weinberg claims that "it is not clear whether the universal and historical elements in our sciences will forever remain distinct" (p. 34). Then he also takes the view that "the most extreme hope for science is that we will be able to trace the explanations of natural phenomena to final laws *and* historical accidents" (p. 37). To crown this inconsistency, he dangles before his readers' eyes the possibility that "what we *now* regard as universal laws will eventually turn out to represent historical accidents" (p. 38). The reader would have profited had Weinberg spoken of irreducible or apparently irreducible specificities instead of historical accidents.

But if Weinberg had been willing to make such clarifications, he would have had considerable difficulty in selling in the same context the ideas of a sub-universe and a mega-universe. What he means is that the specificities of our universe might be the probable results of the varieties inherent in a mega-universe ruled by universal laws. He does not consider the question: Why we should not take the mega-universe too for a statistical subclass of a super-mega-universe? He does not consider the question: Can statistics be a first or a last step either in logic or in ontology? Throughout the book Weinberg is reluctant to confront such basic questions that should seem to stare in the face anyone desirous of a modest measure of logic.

Dreaming about a final theory can indeed be an exercise that can be universally destructive. A curious remark of Weinberg's in the same context makes this all too clear. He argues that it is very difficult to draw the line beyond which the final theory would cease to be competent. He could, of course, argue that in its basics chemistry has become a part of physics. With less convincingness, one could even claim that molecular biology is a branch of physics, whatever the difficulties of giving a reductionist explanation of entire living organisms. It is also possible to imagine that brain research will eventually identify "some physical system for processing information that corresponds to our experience of consciousness itself." It is at that point that Weinberg most plainly reveals his philosophical poverty: "That may not be an explanation of consciousness, but it will be pretty close" (p. 45). Actually, it is at that stage that the difference between the act of consciousness and its physico-chemical equivalent should seem to loom most staggering, philosophically that is. But it takes a physicist of the stature of Einstein to recall, and approvingly so, the saying that "the man of science is a poor philosopher."

The way Weinberg analyzes the notion of reductionism illustrates this poverty. Among the various meanings of reductionism listed by him, one would look in vain for the culturally crucial one. It is the kind of reductionism whose seeds were planted by Descartes, Hobbes, Spinoza, and Hume, turned into a Weltanschauung through 19th-century materialism, and taken for a supreme ideology in 20th-century scientism. According to that kind of reductionism a scientifi-cally ordered knowledge is the only valid form of reasoned cogitation.

The entire thrust of Weinberg's diction is an endorsement of that reductionism. Many evidences of this can be found in chapters 4-8, or his account of the steps made in this century toward a final theory. First comes, of course, quantum mechanics, which, according to Weinberg, will have to be a part of the final theory. This may very well be so, but only at a price which Weinberg does not really notice. For him it is enough that quantum mechanics works. But contrary to what he claims, this is not a realist position. The imaginary debate he formulates between a Scrooge and a Tiny Tim is not between realism and positivism, but between a grim resolve not to face up to questions of ontology and a rather poorly articulated operationism. It is that grim resolve which is the essence of the Copenhagen interpretation and not, as Weinberg would have it, "the sharp separation between the system itself and the apparatus used to measure its configuration" (p. 74).

In sharing that resolve Weinberg's relies on esthetics, as he analyzes three major interplays of theory and experiment as steps toward the final theory. Aesthetics, he claims, was the major support on behalf of general relativity, on the quantum electrodynamics and the theory of weak nuclear force. The upshot is that one should not look for a science of science, but merely for some recondite beauty which transpires in the "description of the sort of behavior that historically led to scientific progress—an art of science" (p. 131).

A historian of physics may not find it fruitful to search for a pervasive evidence for that behavior. A logician, however, can put his teeth into Weinberg's declaration that the beauty in question is "the beauty of perfect structure, the beauty of everything fitting together, of nothing being changeable, of logical rigidity" (p. 149). It then follows that "if we ask why the world is the way it is and then ask why that answer is the way it is, at the end of this chain of explanations we shall find a few simple principles of compelling beauty" (p. 165). Here Weinberg once more tantalizes his reader with the suggestion that not only are physicists moving in the right direction but are perhaps not so far from the goal.

Only those who take philosophy for some foggy mental exercise will be surprised on finding Weinberg disagree with the view that philosophy may be a potential help for physicists. Unfortunately, he considers to some extent only positivism, relativism, and subjectivism. He fails to see that a genuinely realist epistemology and metaphysics formed the matrix of Western science, whose universally valid objectivity he firmly upholds as one of three values all mankind can share. The two others are democracy and contrapuntal music. A rather narrow prospect for mankind.

It is in chapter 8 that Weinberg confronts specific obstacles in the way of formulating a final theory. Tellingly those obstacles are the kind of specificities he finds difficult to countenance. He calls them arbitrary features that include "a menu of particles, a number of constants such as ratios of masses, and even the symmetries themselves" (p. 192). These obviously unwelcome features are parts of the standard model of elementary particles and of three forces: electromagnetic, weak and strong nuclear forces. But the model has nothing to offer about gravitation. A reason for this is that only at extremely high energies does the gravitational force become equal to the other forces between two typical elementary particles.

To cope with the special status of the gravitational force, usually referred to as the hierarchy problem, some sort of supersymmetry may be needed which in turn would imply the existence of further Higgs particles. Yet the same consideration also shows that only at energies a million billion times higher than the highest energy reached in today's accelerators would all the four forces appear united. In other words, the SSC, which would exceed by a factor of ten the highest energy available through CERN, would still be enormously removed from providing experimental evidence on behalf of Final Theory, identical as it may or may not be with TOE, or the Theory Of Everything. The best one could expect from the SSC would be but a very distant pointer toward that ultimate goal. Or as Weinberg admits: "The particle physicists' campaign for the Super Collider has been spurred by a sense of desperation, that only with the data from such an accelerator can we be sure that our work will continue" (p. 210).

The work in question is obviously the verification of theories by experiment, the touchstone of truth of anything in physics insofar as it has to do with the real world. Not that Weinberg would spell out bluntly what Einstein had remarked around 1920 concerning the observational verification of the general theory of relativity. Were only one of the predictions of the theory to be refuted by experiments, Einstein warned, the entire theory would irremediably be shattered. In fact, Weinberg is too much concerned about the perfection of the theory itself which he sees prefigured in the string theory. But he admits that "at present we have no criterion that would allow us to tell *why* that string theory is the one that applies to the real world." What he adds in the same breath reveals even more his real aspiration: "Once again I repeat: the aim of physics at its most fundamental level is not just to describe the world but to explain why is it the way it is" (p. 219). That such an aim pertains not so much to physics but to philosophy or logic becomes clear a few pages later as he claims that even the simplicity of the symmetry of supersymmetry ought to be explained.

It is in chapter 10, entitled "Facing Finality," that Weinberg comes to grip with the truly fundamental questions raised by a Final Theory. He begins by taking issue with Popper in particular who argued against finality in science. (Weinberg fails to use the principal objection against Popper, or the charge that Popper is guilty of self-contradiction as he argues that every proposition with a truth content ought to be falsifiable, except this very proposition. Hardly a minor oversight on the part of a philosopher so keen on logic.) Weinberg rather sets great store by the historical convergence of physical theories.

But from this it is not possible to go on to what he calls, by echoing Robert Nozick, the principle of fecundity. According to that principle, all logically possible constructs ought to have a physical existence, our universe being one of them. Weinberg knows that the principle of fecundity is not verifiable because all such constructs (and corresponding universes) are strictly isolated from one another. But taken in its isolation, the set of conceptual constructs that give rise to our physical universe is "so rigid that there is no way to modify it by a small amount without the theory leading to logical absurdities." But if it is the concepts that create reality then it follows not only that "in a logically isolated theory every constant of nature could be calculated from first principles" but also that "a small change in the value of any constant would destroy the consistency of the theory" (p. 236).

This is the second time that the word "consistency" occurs in a crucial context without Weinberg's noticing the cruciality of it, in spite of his making a crucial distinction. The final theory, he writes, would be "like a piece of fine porcelain that cannot be warped without shattering." This means the crucial difference that "although we may still not know why the final theory is true, we would know on the basis of pure mathematics and logic why the truth is not slightly different" (pp. 236-37).

Whether mere conceptual constructs would construct physical reality is a problem that has not ceased plaguing Platonists (even those whom Weinberg calls moderate Platonists among whom he classes himself). Here it should be enough to focus on Weinberg's way of coping with the problem of knowing whether the final theory is true. His answer contains more than what meets the eye: "The problem seems to be that we are trying to be logical about a question that is not really susceptible to logical argument: the question of what should or should not engage our sense of wonder" (p. 238).

If the reader has been swayed by Weinberg's heavy reliance on the role of esthetics in doing good physics, he will not see a most disturbing fly in the ointment. How, one may ask, is it possible, that in a book where hundreds of names fill the index, one looks in vain for the name of Kurt Gödel. All the more so as it is now over half a century since the world of mathematics was shaken to its very foundation by the publication of Gödel's incompleteness theorems. According to them no non-trivial set of mathematical propositions can have its proof of consistency within the system itself. The reader of this review can easily note some earlier warnings about the bearing of Weinberg's repeated, though inattentive, uses of the word consistency.

The impossibility of formulating a final theory about the physical world, which would contain its own proof of being true, relates not to the always shifting grounds of aesthetic considerations but to logic or mathematics. The reason is as straightforward as is the highly mathematical character of every significant physical theory. That leading physicists still have to become aware of this elementary inference (and indeed of the long shadow which Gödel casts not only on mathematics but also on physics) may be perplexing enough. The present reviewer is certainly perplexed both for a general and for a particular reason.

The general reason relates to the fact that since 1966, when I first set forth in my book, *The Relevance of Physics* (University of Chicago Press) the bearing of Gödel's theorems on physics, I have re-articulated the same point in at least a dozen books (including my Gifford Lectures, *The Road of Science and the Ways to God* and my Farmington Lectures, Oxford, *God and the Cosmologists*. But it seems that within a certain circle of physicists only publications written by members of that circle are being read and discussed.

The other reason is very specific. In 1976 I shared the same panel with Prof. Weinberg (as well as four other physicists: Gell-Mann, Weisskopf, Hoyle, and the Harvard philosopher Hilary Putnam) at the Nobel conference on cosmology and fundamental particle physics. There all the panel members and a two-thousand-strong audience heard Prof. Gell-Mann declare that within three months but certainly within three years he would be able to explain why the standard model has to be necessarily true. When it was my turn to make a comment on Prof. Gell-Mann's presentation, I merely said that he would not succeed, a comment that markedly displeased him. On his asking me for some explanation, I referred to Gödel's theorems, theorems with which he admittedly was not familiar.

Later Prof. Putnam told me that he and Prof. Weinberg returned to Harvard on the same plane and that part of their conversation related to the meaning and bearing of Gödel's theorems. About the same time I was giving a lecture on final theory at Boston University. Following my presentation, in which much emphasis was laid on Gödel's theorems, somebody from the audience came up to me to chide me for apparent plagiarism. A week or two earlier, he said, he had attended a lecture by Prof. Gell-Mann, who, with a reference to Gödel's theorems, registered the impossibility of constructing a Final Theory. I restricted my reply merely to telling him what had taken place at that Nobel conference.

Clearly, one need not bring in the always slippery fish of esthetics when plain mathematical logic will suffice. Apart from that logic, those who find only in a personal Creator the ultimate reason for the existence of a specifically ordered physical universe, have always known that a necessarily true final theory is a philosophical pipedream. But such a way out of the labyrinths of a search for a final theory is not open to Prof. Weinberg, who devoted to God the next-to-last chapter of his at times very entertaining and informative dreaming about the final theory. But the God he is ready to consider is not a Creator who has the freedom to bring into existence any of an infinite number of possible worlds. What he endorses is panentheism which, apart from its impotence for coping with the moral evil in the world, cannot even justify the enduring question of perennial philosophy about the physical world: why such and not something else?

But within panentheism one cannot even come to grips with the comparatively trivial question of why the human mind can err, of why it can remain closed to the obvious, and—last but not least—of why some Honorable gentlemen can do brazenly stupid things after having been instructed by our most brilliant physicists. Of course, the latter would not have told them about what is, by the same stroke, the good news and bad news for physicists.

I first put the matter in this form at the beginning of a talk I was invited to give at the Batavia accelerator in April 1992 on the last word in physics. The bad news is that no physical theory, however sweeping and successful, can be considered final as long as Gödel's theorems remain valid, insofar as the theory is non-trivially mathematical. But precisely because of this, and this is the good news, physicists can never work themselves out of a job. Such is a rather pragmatic motivation compared with the far more superior urge to try to find out ever more about the physical universe, which does not fail to show itself supremely, though also most specifically, ordered in a quantitative way. Gödel's theorems or not, it is not scientific to assume that sometime in the future no unexpected features will turn up as experiments (that ultimately rule physics) go on and play havoc with theories, including the one held to be final. One more reason to wake up from certain dreams, finally.

12

Science and Religion
in Identity Crisis

Three hundred or so years ago not a few scientists spoke of science and religion as united in a holy alliance. Two hundred years later theologians could do little about the warfare in which science and religion appeared to be locked forever. Today, many theologians and some scientists speak of the mutual integration of science and religion, these two paramount forces in human life.

From both sides all too often mere generalities rather than tangible specifics are offered. This may already indicate a lack of simultaneous competence in both fields. Actually, those generalities suggest that an identity crisis may be enveloping both science and religion, and to an extent far greater than one may suspect.

The religious side of that crisis is easier to diagnose by a mere look at programs of instruction offered in most departments of religion and religious studies as well as divinity schools and theological faculties. Actual exposure to what goes on in those places can

Lecture delivered on April 2, 1991, as part of its program on Science, Technology and Religious Ideas, University of Kentucky, Lexington.

readily bring into focus a feature typical of most of them. Whenever a question is posed, only a multiplicity of answers istolerated. Even the slightest effort to cut through that multiplicity, within which contradictory stances too are acceptable, is frowned upon as judgmental. The result is the rise of that church where, to paraphrase a remark of Chesterton, each communicant is sharing the other's unbelief.

A biting portrayal of this pathetic situation was given less than a year ago in a book, *The Search for God at Harvard*, written by Ari L. Goldman, religion reporter for *The New York Times*. It may not have been a sound idea at all on Mr. Goldman's part to spend a full year at Harvard Divinity School to search for God there. Actually, the true target of Mr. Goldman's search was not so much God as some experience about Him. Such a search could, of course, have ended, even if successful, only in mistaking God's identity for some religious experience with no real identity.

The variety of religious experiences to which Mr. Goldman found himself exposed in that prestigious divinity school seemed to serve the purpose of concealing their true identity. Nothing has indeed changed there since William James, that legendary guru of religion as "experienced," came up with his own theory about the varieties of religious experience.

Had Mr. Goldman thought, while at Harvard, of William James, he would not have been forced to identify the Christian religious experience as "the most elusive experience" of his early days in the "Div School," as it is called there in a quasi-affectionate tone: "If, for example, there was a mention in class of the divinity of Jesus, the lecturer would offer an apology to the non-Christians in the room." No wonder Mr. Goldman found shattered his expectation of encountering some religious experience which could br identified as "old-type Christian piety." This piety has always been rooted in clearly identifiable dogmas, but the Div School's atmosphere was "religious relativity." There "religious truth did not seem to exist" at all.[1] What Mr. Goldman could not find in the classrooms of the "Div School," he also failed to find in its imposing Chapel. Whether during the daily Noon service, which he faithfully attended, or whenever he peeked into the Chapel in going to classes in the morning, he "never saw anyone on his or her knees." At most, he saw, though not frequently, "someone sitting there meditating."[2]

Clearly, the "Div School" at Harvard did not advance a whit beyond the state of affairs which set the tone at Yale too, as searingly portrayed a generation ago in *God and Man at Yale*. Of course, that portrayal was possible only because its author W. F. Buckley offered an evaluation in terms of definite values, or standards. Whether these are called dogmas or not should seem irrelevant. If Mr. Goldman held any dogma, it was his visceral attachment to orthodox Jewish practices, although he did not care to put it on clearly identifiable intellectual foundations.

This is why he was torn about Roman Catholicism. On the one hand, he felt deeply attracted to the Mass. On the other hand, he could not warm up to dogmatic Catholicism. It is difficult to decide whether he deplored the present status of Catholicism, as he perceived it. Although he seemed to be upset over the Catholic Church's loss of moral authority within society, he was ambivalent about its cause, "the internecine struggles over authority with Rome and the anti-abortion cause."[3]

Only if one is wholly unfamiliar with the long-standing uncertainty of Congregationalists about their own identity can one voice surprise at the utterly elusive identity of religion in a Divinity School and University with Congregationalist roots. The doctrinal atmosphere at Harvard Divinity School reminded Mr. Goldman of that nutshell summary of liberal Protestantism which H. R. Niebuhr had given half a century ago: "A God without wrath brought men without sin into a kingdom without judgment through the ministrations of a Christ without a cross."[4] Such a religion could not be distinguished from Mr. Goldman's Judaism, to say nothing of reformed and conservative Judaism, except for his attachment to ritual laws. In all those denominations, one can freely wear the p.c. (politically correct) badge, this most effective sedative against the pressing need for true identification.

A p.c. religion will have no problem being integrated with science, though the relation may not amount to more than a convenient cohabitation that can be initiated, acted out, terminated, resumed, and reinterpreted on short notice. Cohabitation is always a dissimulation of the true identity of a rapport, an identity crisis in short. The religious side of that cohabitation can only function as religious syncretism. Thus no real difference will be claimed between nature worship and a worship steeped in that supernatural which is a Creator free to create or not to create what is called Nature writ large, that is, a universe.

Within that syncretism every form of religion can be accommodated. There polytheism, with its worship of idols, will not appear too distant even from a worship that forbids the making of graven images of God. And when God and nature are fused to the extent in which this is done in pantheism, not only can one's religion not be identified, but even one's own identity diminishes to the vanishing point. In no form of pantheism has there ever been a place for that personal immortal soul which alone makes one's identity (and one's religion) meaningful and raises it above the lowlands of mere aestheticism.

Syncretism, or the abolition of true identity, certainly foments heavy reliance on verbalism, which is in view, for instance, when pantheism is promoted in the guise of panentheism, or the idea, by itself perfectly orthodox, that God is everywhere and in everything. Syncretism, or religion's identity crisis, is all too often couched in such noble words as ecumenism, global consciousness, and moral rearmament, to say nothing of such dubious labels as Gaia and New Age.

The so-called "mere Christianity," first proposed around 1675 by William Baxter, a Puritan divine tired of religious controversies,[5] was a symptom of identity crisis. The symptom resurfaced when in 1943 C. S. Lewis resurrected Baxter's idea in a book, *Mere Christianity*, that made religious history for the latter half of this century of ours. C. S. Lewis could offer but his gut-feeling as to what that mere Christianity was when he said that he meant by Christian faith that "which is what it is and was what it was long before I was born and whether I like it or not."[6] No more clarity was shed on the subject by his equally elusive definition of Christian belief as the one "that has been common to nearly all Christians at all times."[7]

This is not to suggest that C. S. Lewis was not aware of the problem of leaving out of "mere Christianity" all items smacking of controversy in order to focus attention on items non-controversial. But were there such items or tenets? No less importantly, even if there were some, could they be discoursed upon for any length of time without bringing up matters not only controversial but also pivotal for the articulation and defense of points commonly held by almost all?

That almost complete unanimity evaporates like the morning mist once one raises the question of whether "mere" Christianity implied miracles. Without talking of miracles in *Mere Christianity*, C. S. Lewis held miracles to be an integral part of the Christian proposition. Otherwise he would not have defended the possibility, as well as the reality, of miracles against Humeans, erstwhile and modern.[8] Yet, many Christians who hold high a Christianity restricted to its mere basics refuse to face up to non-biblical miracles that are much closer to us and far easier to investigate. Nor do they see that agnosticism about post-Biblical miracles is destructive of faith in biblical miracles as well.[9] Still other Christians prefer not to speak of miracles precisely because they want to integrate science and religion, though on terms dictated by the interpretation of science given by most scientists, who have no use for miracles at all.

Some defenders of biblical miracles grant them only because they take the space-time of relativity and the indeterminacy of quantum mechanics for a scientific possibility of miracles. They do not realize that in doing so they do away with miracles as well as with the physics of relativity, which is based on the strict continuity of the physical and for which time is, as Einstein once memorably admitted, a mere parameter of measurements.[10] It seems that they are not sure of themselves and of the grounds, not at all scientific but deeply philosophical, on the basis of which alone one can confidently speak of miracles, and do so in a healthy disregard for science.

It is not even certain that today Christians are unanimous in their belief in creation out of nothing and in time. Christians are a motley lot in their interpretation of Genesis 1. Stances vary from taking it for the relic of *pre*literary legends to interpreting it in a grimly literalist sense, as is done in creation science.[11] Typically, proponents of creation science make much, as if to cover their own uncertainties, of the uncertainty of various scientific conclusions and of the uncertainty of scientists about them.

Last but not least, there is Christianity's central belief: redemption in Christ. Newborn Christians all too often do not want to hear about the conceptual certainties of dogmatic definitions achieved in the great christological debates of the early Church. Even within the proverbially dogmatic Catholic Church, discourses can be heard about Christ that can only prompt any consistent thinker to part with belief in Christ as the only begotten Son of God. Quite recently an Anglican bishopric was awarded to the Regius professor of divinity in Oxford whose scholarly reputation rests on a programmatic rehabilitation of Arius.[12]

Even Arius held Christ in a much higher esteem than the Roman Catholic authors of two recent lives of Christ. According to one of them, who protests his *Roman* Catholic orthodoxy, Jesus *may* have been a celibate, but he certainly had real brothers and sisters. But that author's religious identity becomes doubtful when he argues that, since historical scholarship proves only that Jesus was a peasant Jew, his divinity can only be had on faith.[13] He seems to be blissfully unaware of the historic fact that this dichotomy between historic or "scientific" evidence and faith has been a chief source of depriving all too many Christian professions of faith of the identity they should have manifested. The very identity of Christ is indeed in widespread doubt among Christian theologians who lend receptive ears to some "experts" on the Qumran scrolls who maintain that Christianity in its earliest form contained nothing essentially new that Judaism at that time did not contain.[14]

The systematic leveling of Christ to very low human levels is, of course, part of the carefully cultivated uncertainty about sin. Genuine Christian awareness of the reality and seriousness of sin has now for decades been under mounting pressure to align itself with "new" perceptions about sin. One is the view that sin is a psychological infirmity; another consists in not perceiving sin at all. For fashionable thinking there are only so many patterns of behavior. Consequently, any behavior, once it becomes a pattern, that is, once it is acted out by a statistically significant number of people, can claim social acceptance. From there it is but a short step to claims for legal protection and moral respectability, as if legal were equivalent to moral.[15]

Christian (and Jewish) awareness seems to have come a long way from the injunction given in the book of Exodus: "You shall not repeat a false report . . . . Neither shall you allege the example of the many as an excuse for doing wrong" (Ex 23:1-12). Too many Christians seem to be worlds removed from those forebears of theirs who accepted cheerfully the truth of the words: "You will be under pressure in this world," because "no disciple is greater than his master." Christian theologians are few and far between who have a style with at least a touch of Tertullian's incisiveness: "Christ said, 'I am the truth.' He did not say, 'I am the custom.'"[16] Far many more are those theologians whose "reasonings" illustrate Edmund Burke's acid remark: "Custom reconciles us to everything."[17]

Speaking of Exodus is as good a reason as any to say a few words about the disarray in which Jews of our times find themselves concerning their religion. Varieties of opinion range from literally militant Zionism to no religion at all. I wonder whether those Christians who firmly believe in the personal immortality of their souls and in subsequent resurrection would find much sympathy among Jews, except for the most orthodox. Cultural conservatism, very popular among American Jews, is not enough. It does not by itself lead beyond this life on earth to an otherworldly compensation for cruel deprivations suffered in this life. That there can be no such compensation was the gist of Norman Podhorotz's objection after I had spoken in Moscow, in June 1989, before a meeting sponsored by the Soviet Academy of Sciences and Moscow University, on belief in the existence of God as supported by science. In that speech[18] I also referred to that belief as the only ground on which one can think of an eventual compensation for tragedies whose number, in this life, is far more than legion. And yet I was speaking only of the hundreds of millions of innocent sufferers without saying a word about those—not a few—who never suffer in this life any punishment for horrible crimes.

About two other religions, Islam and Buddhism, both increasingly popular in the West, a few remarks should suffice about their possible relation to science. Behind the Islamic religious revival lies a feature hardly ever noted by Western observers. Owing to the demands of technologization, Islamic lands now have to provide scientific education on a large scale. This in turn confronts their people with the challenges of empiricism, rationalism, and positivism. The result is a turmoil, for the moment still largely under the surface, but a turmoil which the Islamic clergy seems to perceive in its real magnitude. There will be a replay, but in a much more dramatic form, of the famed medieval debate between the Muslims who are genuine mystics, and those who are rationalists and therefore are Muslims only in name.

A millennium ago, leading Muslim mystics such as al-Ashari and al-Ghazzali held that reference to the laws of nature was a blasphemy against Allah's omnipotence. The leading Muslim rationalists, such as Ibn-Sina (Avicenna) and Ibn-Rushd (Averroes), held that the truly enlightened Muslim can hold the idea of triple truth: The lowest truth, or the truth of catechism, was for the populace; the middle level was for the imams, repeating the Koran; the highest form of truth, the truth of science, was the privilege of the enlightened. These, however, were to keep that third truth to themselves, lest they suffer the consequences.[19] The medieval Muslim world could not find a middle ground, such as the one worked out by the great scholastics shortly afterwards concerning faith and reason. Whether modern Muslim intellectuals can work out a satisfactory balance remains to be seen. But they had better recognize that an identity crisis is looming large over their heads, a crisis fueled by science, or rather by the impossibility of living today for more than a few minutes without taking advantage of this or that scientific tool.

As to Buddhism, its numerous varieties may in themselves suggest a chronic identity crisis. Such a crisis has plagued many of those young people in the West prior to their embracing Buddhism as a religion. At any rate, in its classical method, aimed at giving an escape from the self, Buddhism can hardly escape the suspicion that it offers a cure for an identity crisis by depriving the self of its identity. In fact, in all the great philosophical presentations of Buddhism, as also in its amalgamation with Confucianism and Taoism, a central place is occupied by the denial of what is known in Western logic as the principle of identity and non-contradiction.[20] No wonder that Buddhism has received a very sympathetic consideration by those who see in the Copenhagen interpretation of quantum mechanics the highest form of enlightenment because that interpretation is taken as the equivalent to a physics which too many take for the ultimate in science. To make matters culturally far worse, Niels Bohr lent his full scientific prestige to establish the principle of complementarity, which is an integral part of that interpretation, as a philosophy that would give better guidance in life than any religion ever could.[21]

Dime a dozen are the books and articles written by theologians and scientists, at times together, in which one finds it registered with great satisfaction that science and religion are complementary like the particle and wave aspects of matter.[22] This claim certainly deserves some scrutiny even within the perspective of this lecture. The claim leaves in studied indefiniteness the religion in question. A moderately careful reading of the philosophical works of Bohr, Schrödinger, Max Born, and Heisenberg should easily reveal a conviction common to all of them. They would indeed protest to a man against the claim that the principle of complementarity entitles one to make a rational plea for immortality, to say nothing of revelation, miracles, resurrection, and a judgment pertaining to an eternal reward or eternal punishment.[23] The principle of complementarity tolerates only a religion which is reduced to that sheer aestheticism where one is faced with ever shifting moods and styles, and with a perennial craving for fulfillment that never comes.

But the principle of complementarity, as taken for a philo-sophical and religious panacea, or cure-all, leaves even science, that is, the science of quantum mechanics, in an identity crisis. A priceless glimpse of this was provided by no less an insider than the late Professor Dirac. It tells something of the measure of that identity crisis that what Dirac said in the broadest scientific daylight, the Jerusalem Centennial Einstein Conference in 1979, has been studiedly ignored by the scientific establishment, and by its chief ally, or perhaps millstone around its neck, the establishment known as the philosophers and historians of science. Dirac said nothing less at that Conference than that quantum mechanics, as it stands today, will have to be reformulated along the lines of strict predictability as demanded by Einstein: "I think that it is very likely, or at any rate quite possible that in the long run Einstein will turn out to be correct even though for the time being physicists have to accept the Bohr probability interpretation—especially if they have examinations in front of them."[24]

Clearly, this is not the kind of diagnosis which its subjects would sedulously recall to themselves, let alone to their students. Ostriches love to bury their heads in the sand lest they should be forced to face the true identity of their predicament. This is precisely what the champions of the Copenhagen interpretation of quantum mechanics have done with respect to its most considered appraisal by J. S. Bell, of Bell's Theorem fame. On the one hand, they have not ceased to recall his famed theorem, which they take for the final proof that ultimately everything is haphazard. Yet all he did was to show that quantum statistics is more successful than classical statistics in coping with a certain kind of coincidence in radioactive emission. Two years later, on finding subtle illogicalities in the "reduction of wave packets," a pivotal issue in theoretical quantum mechanics, he felt impelled to conclude that quantum mechanics "carries in itself its seeds of destruction."[25] Physicists, and philoso-phers of physics, still have to pay adequate attention to this much more profound conclusion of Bell. When they bring it up, they usually display that touch of nervousness such as transpires from K. Gottfried's very polite rebuttal of Bell's claim.[26] Theologians who continue integrating their field with quantum mechanics should pay heed. For they may be unwittingly promoting their own self-destruc-tion, the ultimate form of identity crisis.

Much has been said about the identity crisis that beset physics toward the end of the 19th century. Much less spoken of are the early traces of that crisis. There is much more than meets the eye in an apparently innocent facet of Newton's *Principia*. It contains not a single paragraph on the philosophico-methodological questions raised by the fact that he had written not merely a *Principia*, in itself a grave word, but a *Philosophiae naturalis principia mathematica*. One would in vain interrogate Newton as to what he meant by nature, by philosophy, and even by mathematics.

Newton may not even have been absolutely sure of himself as a physicist. Otherwise, he would not have spent precious hours in erasing from his manuscripts references to Descartes, lest posterity should suspect that he owed something to the Frenchman. No wonder. Cartesians in France greeted the *Principia* with the remark that for all its fine qualities, it was not physics.[27] What was it? It certainly was not a mechanistic physics insofar as this means mechanistic models, of which the *Principia* was conspicuously void. Only some time later were those models grafted, and in large numbers, on the mathematics of the *Principia*, which ushered in the age of infinitesimal calculus.

Although a rigorous proof of the "limit" did not come until the early 19th century, Newton's infinitesimal calculus stood for a blissful state of total certainty, with no trace of identity crisis. The next scientifically momentous book with the word *Principia* in its title was published in 1910, under the title, *Principia mathematica*.[28] It did not contain a retraction by its co-author, Bertrand Russell, concerning his earlier description of mathematics "as the subject in which we never know what we are talking about, nor whether what we are saying is true."[29] Later, Bertrand Russell qualified his agnosticism about mathematics only to the extent of saying that physics is "mathematical not because we know so much about the physical world, but because we know so little; it is only its mathematical properties that we can discover."[30]

This warning of Bertrand Russell's should be ample food for thought for those who cultivate other branches of physical science, especially the various branches of life-sciences. Being the most exact among all empirical investigations, physics is eagerly imitated by other empirical sciences. Their cultivators may suffer some identity crisis owing to their being overshadowed by their glamorous big brother, big physics. In addition, this remark, if pondered enough, may even bring back to their senses various cultivators of the humanities. For whenever psychologists, sociologists, or historians try to imitate physics, and quite a few of them do, they merely reveal symptoms of identity crisis.[31]

This is even truer of philosophers, to say nothing of theologians. The program of a so-called scientific philosophy is a fairly old fad that had such prominent devotees as Descartes, Hobbes, Hume, Kant, and Comte, and produced long-discredited systems of philosophy. The fad of a "scientific" theology is rather new. It produced lengthy treatises about the theology of the world, without a paragraph in them about that world which is the universe.[32] One could also refer to books with "insight" and "method" in their titles, though with no clear guidance in their contents as to what qualifies for insight and in what true method consists.[33] If only such theologians aimed at nothing more than offering a well-argued rational discourse! Neither philosophy nor theology is or can be a science in the way in which physics is one, but both should be eminently rational, that is, well-reasoned, instead of being enveloped in endless chains of vague metaphors and new-fangled buzzwords.

Were theologians to handle their own field as a discourse that demands utmost attention to clarity and consistency, they would discover a curious thing about science and scientists, and in particular about physics. Until about a hundred years ago, physicists could be seen from the outside as safely entrenched in their certainties. James Clerk Maxwell well characterized that state of affairs when in 1870 he described the Royal Society as "the company of those men who, aspiring to noble ends . . . have risen above the regions of storms into a clearer atmosphere, where there is no misrepresentation of opinion, nor ambiguity of expression, but where one mind comes into close contact with another at the point where both approach nearest to the truth."[34]

A quarter of a century later, Maxwell's electromagnetic theory appeared to have a truth content which resembled Nicolas of Cusa's definition of the universe: its center everywhere, its circumference nowhere. For this is the unintended gist of Hertz's famed remark: "Maxwell's theory is Maxwell's equations."[35] In other words, nobody really knew what was the true identity of those equations and that theory. Hertz's definition implied that circularity indicative of an identity crisis. Today, in this age of string-theory, zero-point oscillations, embryo universes and what not, it is even more difficult to specify the identity relation of each and every part of very successful physical theories with physical reality.

But the true identity crisis of science lies elsewhere. This is not to suggest that in empirical sciences, even outside physics, one could not find momentous traces of identity crisis even today. Evolutionary biology still has to come to terms with two rude jolts that knocked many of its cultivators out of their blissful state of mind. One was the realization, quite recently, that the theory of the origin of life as proposed by S. L. Miller in 1952 cannot cope with the high temperatures which certainly prevailed on the primitive earth three to four billion years ago.[36] The other is the evidence of major catastrophes to which the earth is exposed every 26 million or so years, some of which result in substantial extinctions of all life forms.[37] The identity crisis that resulted from this is well exemplified by the so-called theory of punctuated equilibrium of evolutionary development. At a national gathering of its devotees the level of consistency was such as to prompt one participant to remark that he would find more intellectual honesty at a national gathering of second-hand car dealers.[38]

The real source of identity crisis in science today does not lie with its internal problems of which it will very likely never run short. For who can give assurance that future decades and centuries hold no surprising discoveries, totally unimaginable today? The real source of that identity crisis lies in the fact that scientists have, to a considerable degree, ceased being the official spokesmen of what science is about. The image of science entertained by society at large, and even within academic circles, is now determined as much, if not more, by what philosophers and historians of science, and mere science writers, state about science.

The change should seem enormous. A hundred or so years ago, relatively little was written about the philosophy of science. Authors of such books were usually either philosophers with very little training in science, or scientists with as little familiarity with philosophy. William Whewell, in the 1830s, was the first prominent scientist to write a serious book on the philosophy of science, and he was not imitated until Ernst Mach, Pierre Duhem, and Henri Poincaré came along around 1900 or so.

From the 1920s there was a rapid increase in the number of prominent scientists who wrote books on the philosophy of science. To speak only of physicists one may recall the names of James Jeans, Eddington, Bohr, Born, De Broglie, Margenau, Whittaker, Schrödinger, Heisenberg and, more recently, Feynman. But the books that really formed the "standard" image of science were written by philosophers, at times with proper training in physics, at times with no such training. Moritz Schlick was not a physicist, nor were Karl Popper, Hans Reichenbach, and Herbert Feigl. While Thomas Kuhn had full training in physics, this was not true of Lakatos and Feyerabend. Yet it was these who imposed on our culture the view that science is intrinsically uncertain about itself. What can such an uncertainty breed if not an identity crisis, and a chronic one?

No major articles on science contributed by philosophers of science to major encyclopedias will give a fair certainty as to what science really is. There are still some grim inductivists around, as well as some dreamy-eyed Platonizers. Classical positivists, in the style of Auguste Comte, are few and far between, but the number of Machists is great, though few of them take consistently Mach's sensationism which ultimately led him to embrace Buddhism.[39] Most would endorse some form of the idea of science as based on the hypothetico-deductive model. Yet, divergences of opinion abound as soon as one comes to the art of forming hypotheses and to the legitimacy of deductions. With the exception of Duhem, who was a realist and held high common sense as the only consistent starting point,[40] they would have only scorn for T. H. Huxley's famous definition of science as "trained and organized common sense."[41]

Many philosophers of science offer explanations of science that border on being nonsensical. Such is the case when science is taken for an exercise in falsification. But, if all that science can do is falsify conclusions, not only the identity of its conclusions is at stake but its very own identity as well. Such should seem an inevitable conclusion unless one is denied the liberty to see the contradictory character of the following claim: only such statements as have truth-content are falsifiable, though this truth is immune to the test of being falsifiable.[42] If science is a series of images, of themata, of research programs,[43] what will assure that they fuse into one image, one thema, one program that can be safely identified? That the idea of science as an anarchical enterprise[44] has found devotees may be symptomatic of the identity crisis that breeds anarchists, intellectual and other. That science has also been spoken of as a game, very clever of course, is characteristic of recent decades when too many young academics had it too well and lost sight not only of the identity of their subject but very often of their own self-identity as responsible human beings.

Intellectually very treacherous is the case when science is identified with a big word that everybody uses and nobody defines. Apart from this, it should be obvious that if science is a never-ending chain of revolutions, this very proposition is its own refutation, because the proposition is offered as not being subject to the kind of change which is revolution understood as radical upheaval. Or should we let some philosophers of science have it both ways, namely, to bank heavily on big words and be cagey about what they mean by them? Those who speak constantly of scientific revolutions should come clean whether they mean more than what is implied in the French phrase, "plus ça change, plus ça reste la même chose," which is possibly the best account of all political revolutions.

The phrase has a hardly ever noted philosophical profundity. That depth was covered up when the great apostle of scientific revolutions, T. S. Kuhn, reversed his revolutionary tracks and invoked the principle of essential tension,[45] as if metaphysics could be reinstated through the back door. Yet even by allowing metaphysics back to the stage, Kuhn granted it no more than the role of a mere co-actor, if not an inept complement to the fascinating game of science with quantities.

Yet metaphysics is the very stage which any intellectual performance needs for its being acted out. Even the very word metaphysics indicates that it is not supposed to be something juxtaposed. In that case Aristotle would have called it para-physics and left it aside as something no more worth considering than paralogisms. Aristotle would have felt sympathy for that French phrase, "plus ça change, plus ça reste la même chose." All his philosophy rested on his coming to grips with the problem of identity through change as the safeguard of the sanity of the human intellect.

It is also the safeguard of the sanity of science. Only if changes are such as to leave somehow intact the sameness of things undergoing change does one have a ground to speak of science with no identity crisis. For science deals with things in motion, or change, and has a major stake in the possibility that its observations and conclusions transcend the truth of the moment.

Working scientists would fully sympathize. They are also the ones who somehow sense that all the sophisticated assertions about science in much of modern philosophy of science can but foment and promote the malaise which is the perceived identity crisis of science. It is that malaise which is resented by working scientists who time and again put studies on the history of science on the X-rated lists.[46] History has always been the favorite hunting ground of skeptics and scoffers. They make most of the keen observation of Chesterton that history is so rich in data that one can make "a case for any course of improvement or retrogression."[47]

The identity crisis of science, insofar as it is not turned into an ideology by philosophers and historians of science and by scientists who in their old age wax philosophical, is far less serious than it may appear. The reason for this lies in Bertrand Russell's second remark quoted above. The remark calls attention to a radical limitation of science, of physics in particular. The science of physics knows both enormously much and enormously little about the material world, because it can only know its quantitative properties. Science becomes involved in an identity crisis only when it ignores its own method or when it lets philosophers, eager to promote their agnosticism and subjectivism, take over as the spokesmen for science.

Whereas the cure for the identity crisis of science should seem relatively simple, the cure for the identity crisis in religion is a far more serious matter. At any rate, any theologian who speaks about the relation of science and religion should first come clean as to what religion he stands for. The philosopher must do the same and take the consequences. It seems to me that the art of camouflage, to say nothing of mere chameleonship, is not exempt from the command-ment that forbids lies. Identity crisis can have its cure only in total commitment to the words: truth will make you free. One indeed must first identify and remove a vast heap of débris in order to see the very complex and nuanced truth about the relation of religion and science and about their ongoing interaction.

Interaction makes sense only between too distinct items, factors, or entities. Distinctness in turn can be manifold. Some forms of it impose themselves, some others can make impatient the proverbial system makers who are hellbent on fusing everything into everything and setting the stage for intellectual infernos. Chafe as they may, the fundamental domain of *is,* or plain existence, to say nothing of the all-important domain of *should*, or the domain of moral values and standards, cannot be reduced to the far more entertaining and practically useful domain of purely quantitative relations, the domain of science. The domain of unitary knowledge, insofar as it means a one-track knowledge, belongs to the domain of Utopia on earth, which is not the domain of angels.

The only kind of unitary knowledge about which man can profitably speculate is the knowledge of angels. Unfortunately, latter-day theologians are most reluctant to face up to that topic. Yet by discussing it they would be able to tell modern men (who since Descartes have been trying to play the angel[48]), that mere man must implement his cognitive life in terms of mutually irreducible conceptual domains.

To put it in a perhaps facetious, perhaps slightly blasphemous, but certainly blunt way: domains that God has kept separate for man, man should not try to join together into one single domain. The resulting realm will not be a synthesis, not even a fusion, but a confusion which is all too evident in the manifold symptoms of the real and perceived identity crisis which for some time has been plaguing science as well as religion. I have spoken so much about diagnosis, relatively little about remedies, and almost not at all about a healthy state, because an effective cure depends much on making the diagnosis as full and realistic as possible.

---

[1] A. L. Goldman, *The Search for God at Harvard* (New York: Random House, 1991), p. 43.

[2] Ibid., p. 44.

[3] Ibid., p. 276.

[4] H. Richard Niebuhr, *The Kingdom of God in America* (New York: Harper and Row, 1937), p. 193. In this case too, the ensuing vacuum had to be filled: "For the golden harps of the saints it [liberal Christianity] substituted radios, for angelic wings concrete highways and high-powered cars, and heavenly rest was now called leisure" (p. 196). That was, of course, before the commercially "golden" age of TV and VCR, to say nothing of "grass" and psychedelic religion, had come around.

[5] Baxter's "mere religion" was a platform which already to Baxter's more perceptive contemporaries appeared as one that "might be subscribed by a Papist or a Socinian." See article, "Baxter, Richard," in *Encyclopedia Britannica*, 1991 edition.

[6] C. S. Lewis, *Mere Christianity* (New York: Macmillan, 1952), p. vii.

[7] Quoted in M. Nelson, "C. S. Lewis, Gone but Hardly Forgotten," *The New York Times,* Nov. 22, 1988, p. 27.

[8] C. S. Lewis, *Miracles: A Preliminary Study* (London: G. Bles, 1947).

[9] See my book, *Miracles and Physics* (Front Royal, VA: Christendom Press, 1989. Second revised edition 1999).

[10] Einstein did so under questioning from Bergson, at the Sorbonne, in April 1922. See *Bulletin de la Société française de philosophie* 17 (1922), p. 107.

[11] For further details, see my *Genesis 1 through the Ages* (New York: Wethersfield Institute, 1992), the enlarged text of eight lectures delivered in New York City, under the sponsorship of Wethersfield Institute, in late April and early May, 1992.

[12] R. Williams, *Arius: Heresy and Tradition* (London: Darton Longman and Todd, 1987).

[13] J. P. Meier, *A Marginal Jew* (New York: Doubleday, 1991). The other book, *The Historical Jesus* (New York: Collins, 1991), was written by J. D. Crossan.

[14] One wonders what is to be gained on the Jewish side by the claim, most memorably made in this century by Rabbi J. Klausner, that all ethical and religious tenets of the Gospels occur in early Jewish writings. Whatever the identity crisis that should be logically generated by such a claim, it makes it impossible to identify Judaism as a universal religion. This is what Klausner unwittingly admitted, in addition to destroying his claim that Jesus had offered nothing new in the way of ethics and religion, as he wrote: "Jesus came and thrust aside all the requirements of the national life. . . . In their stead he set up nothing but an ethico-religious system bound up with his conception of the Godhead." *Jesus of Nazareth: His Life, Times and Teaching,* tr. from the original Hebrew by H. Danby (New York: Macmillan, 1926), p. 390. But if one's conception of the Godhead has nothing to do with one's religious and ethical tenets, what is their source? One's national or racial affiliation?

[15] A topic further discussed in my article, "Patterns versus Principles: The Pseudoscientific Roots of Law's Debacle," *The American Journal of Jurisprudence* 38 (1993), pp. 135-57. Reprinted as ch. 1 in my *Patterns or Principles and Other Essays* (Bryn Mawr, PA: Intercollegiate Studies Institute, 1995).

[16] It was with this phrase that Saint Turibius, the famed Archbishop of Lima (1538-1606), countered gold-hungry conquistadors, who tried to justify their evil ways by invoking "tradition." See *Butler's Lives of Saints* (New York: P. J. Kenedy and Sons, 1962), vol. 2, p. 167. The phrase is fully applicable to many captains of capitalism and neo-capitalism.

[17] E. Burke, *A Philosophical Inquiry into the Origin of Our Ideas of the Sublime and Beautiful* (New York: Harper and Brothers, 1844), p. 185 (Pt. IV, sec. xviii).

[18] "Sushchestvnet li Sozdatel?" in *Obshchestvennye nauki Akademiia nauk SSSR* [Moscow] 6 (1990), pp. 170-180.

[19] For details see ch. 9, "Delay in Detour," in my book, *Science and Creation: From Eternal Cycles to an Oscillating Universe* (2d ed.: Edinburgh: Scottish Academic Press, 1987).

[20] As can be seen in such a programmatic identification of quantum mechanics with Eastern philosophies as F. Capra's *The Tao of Physics,* first published in 1975. For a list of unsparing criticisms of Capra's claims, see E. R. Scerri, "Eastern Mysticism and the Alleged Parallels with Physics," *American Journal of Physics* 57 (Aug. 1989), pp. 687-92.

[21] For his words, recorded by a confidant of his, see my *God and the Cosmologists* (Edinburgh: Scottish Academic Press, 1989), p. 221.

[22] For instance, J. Honner, "Unity-in-Difference: Karl Rahner and Niels Bohr," *Theological Studies* 46 (1985), pp. 480-94.

[23] Recent major biographies of Schrödinger, Dirac, and Heisenberg are particularly telling in this respect.

[24] As reported by R. Resnick, a participant at that Conference, in his "Misconceptions about Einstein: His Work and His Views," *Journal of Chemical Education* 52 (1980), p. 860.

[25] J. S. Bell, "The Moral Aspect of Quantum Mechanics" (1966); reprinted in J. S. Bell, *Speakable and Unspeakable in Quantum Mechanics* (Cambridge: Cambridge University Press, 1988), p. 27.

[26] K. Gottfried, "Does Quantum Mechanics Carry the Seeds of Its Own Destruction?" *Physics World* 4 (October 1991), pp. 34-40. Prof. Gottfried discussed the same in a lecture given on February 27th, this year [1991], at Rockefeller University.

[27] Thus the anonymous reviewer of the *Principia* who wrote for the prestigious *Journal des Sçavans* coupled his praise of Newton's work as "most perfect mechanics that we can imagine" with the wish that he would crown it by giving "us a physics as exact as the mechanics." See E. J. Aiton, *The Vortex Theory of Planetary Motion* (London: Macdonald, 1972), p. 114.

[28] By A. N. Whitehead and B. Russell, of course.

[29] B. Russell, "Recent Work on the Principles of Mathematics," *The International Monthly* 4 (1901), p. 84.

[30] B. Russell, *Philosophy* (New York: Norton, 1927), p. 157.

[31] Something akin to the pathological may be on hand when a prominent historian of the antecedents of the American Civil War claims that nothing will be really known about its true cause until the full voting record of the Congress, from the previous thirty years, has been evaluated by a computer. For further details see my forthcoming article, "History of Science and Science in History," *Intercollegiate Review* 28 (Fall 1993), pp. 20-33.

[32] This is particularly true of the so-called transcendental Thomists. For details, see my Père Marquette Lecture 1992, *Universe and Creed* (Milwaukee: Marquette University Press, 1992).

[33] I am, of course, referring to B. Lonergan.

[34] J. C. Maxwell, "Introductory Lecture on Experimental Physics" in his *The Scientific Papers of James Clerk Maxwell,* ed. W. D. Niven (Cambridge: Cambridge University Press, 1890), vol. 2, p. 252.

[35] H. Hertz, *Electric Waves* tr. D. E. Jones (1893; New York: Dover, 1962), p. 21.

[36] In fact, as shown by N. R. Pace, "Origins of Life: Facing up to Physical Setting," *Cell* (May 17, 1991), pp. 531-33, the mere presence of water could be destructive of the process advocated by Miller. See also the report by M. W. Browne about Pace's paper in *The New York Times*, June 18, 1991, section C1.

[37] The first systematic study is the article by D. M. Raup and J. J. Sepkoski Jr, "Periodicity of Extinctions in the Geologic Past," *Proceedings of the National Academy of Sciences* 81 (1984), pp. 801-05.

[38] Reported in *Newsweek* April 8, 1985, p. 80.

[39] For details, see ch. 8, "Mach and Buddhism," in J. T. Blackmore, *Ernst Mach: His Life, Work, and Influence* (Berkeley: University of California Press, 1972).

[40] For further details, see my work, *Uneasy Genius: The Life and Work of Pierre Duhem* (Dordrecht: Martinus Nijhoff, 1984), p. 321.

[41] T. H. Huxley, "On the Educational Value of the Natural History Sciences" (1854), in his *Science and Education. Essays* (London: Macmillan, 1899), p. 45.

[42] The inconsistency echoes the one contained in Comte's memorable dictum: Everything is relative and this is the only absolute truth.

[43] Views promoted by Holton, Lakatos, Elkana and others. For details, see my Gifford Lectures, *The Road of Science and the Ways to God* (Chicago: University of Chicago Press, 1978), pp. 235-36.

[44] Much food for thought is contained in the very titles of some of P. Feyerabend's books, such as *Against Method: Outline of an Anarchistic Theory of Knowledge* (rev. ed.; London: Verso, 1988) and *Farewell to Reason* (London: Verso, 1987). Curiously, he failed to note that any argumentation in support of anarchy has to be non-anarchical in order to make sense.

[45] Tellingly, in his *The Essential Tension* (Chicago: University of Chicago Press, 1977), T. S. Kuhn failed to explain what he meant by metaphysics or even by essence. The book may indeed produce no small tension in any reader of it appreciative of logic.

[46] See S. C. Brush, "Should the History of Science Be Rated X?" *Science* 183 (1974), pp. 1164-72.

[47] G. K. Chesterton, *All Things Considered* (New York: John Lane, 1909), p. 221.

[48] For details, see ch. 1. "Fallen Angel" in my book, *Angels, Apes and Men* (La Salle, IL: Sherwood Sugden, 1983).

13

Science, Culture, and Cult

Science, exact science that is, has become synonymous with the theory of relativity and with quantum mechanics. In the broader cultural context the science of relativity is all too often taken for the proof that everything is relative. This might not have happened if Einstein had followed the suggestion of a friend of his, E. Zschimmer, who in 1922 urged him to rename his theory of relativity as "Invarianten-theorie," or the theory of invariance. In his reply Einstein admitted that the expression "relativity theory is unfortunate and has given occasion to philosophical misunderstandings." Yet he felt that although the new name would "perhaps be better, it would cause confusion to change the generally accepted name after all this time."[1]

Apart from cultural considerations, the new name "theory of invariance" would have done much more justice to the science of

Invited paper for the Plenary meeting of the Pontifical Academy of Science, November 1994. Published in *Science in the Context of Human Culture* (Vatican City State: Pontifical Academy of Science, 1997), pp. 93-118. Reprinted with permission.

relativity. On hearing physicists talk everywhere of "the theory of invariance" the broader public would have come to suspect that Einstein's real achievement consisted in shedding light on some very absolutist aspects of the physical world. Such are the independence of the speed of light both of the velocity of its source and of its detector, and the unchanging form of the basic equations of electromagnetism, regardless of the motion of the co-ordinate system with respect to which they are formulated. The theory of relativity is a form of physical science far more reliably absolutist than Newtonian physics was with its doctrines of absolute space and time.[2]

Also, by 1922, only fifteen years had gone by since the formulation of the special theory of relativity, which at that time was still to become part of the typical physics curricula. In 1922 only a handful of physicists were really familiar with the general theory of relativity, formulated only five years earlier. One of them was Eddington, who gave one of the first and certainly the most readable outline of general relativity, his *Space, Time and Gravitation*, first published in 1920. There he wrote: "The absolute may be defined as a relative which is always the same no matter what it is relative to."[3] Four years later, Max Planck gave a much noted and often reprinted public lecture at the University of Munich under the title, "Vom Relativen zum Absoluten."[4] Still another six years later, no less an expert on general relativity than Willem de Sitter stated in his Lowell Lectures at Harvard University: "The theory of relativity is intended to remove entirely the relative and exhibit the pure absolute."[5]

Planck, Eddington, De Sitter were among the highest scientific authorities of the 1920s. It is safe to assume that they would have readily joined forces with Einstein if he had called for an abandoning of the expression, "relativity theory," as something that posed a potentially great threat to a healthy cultural consciousness. The threat turned out to be most effective as can be seen from the vast literature (philosophical, aesthetic, ethical, political) in which relativism is held high with endless references to the science of relativity.[6] In other words, the science of relativity turned into a cultural anesthetic. It dulled the twentieth-century mind (scientific as well as other) so much as to make it insensitive to the farce which transpires in Auguste Comte's often quoted remark: "Everything is relative and this is the only absolute truth."

It is less profitable to speculate whether the cultural impact of quantum mechanics might have been different from what it turned out to be. Planck, the discoverer of the quantum of action, remained a lonely voice in calling attention to the absolutist character of that quantum.[7] Instead, the magnificent science of quantum mechanics has become a vehicle for the most sinister form of relativization, which is the denial of causality in the name of Heisenberg's uncertainty principle. While relativization in the name of the relativity theory is largely a conceptual game, relativization in terms of the denial of causality strikes its sinister blow at the fundamental or ontological level.

Again, while in the case of relativity, the body scientific failed by omission in its cultural duty, here scientists most actively nurtured and promoted a culturally deadly disease, with Heisenberg in the van. He failed to realize that in the pages of a foremost periodical in physics he made a thoroughly philosophical claim when in 1927 he presented his uncertainty principle and wrote that thereby "the principle of causality has been definitively disproved."[8] The word "definitively" proved only one thing, a thing hardly philosophical. Heisenberg was one of those prominent physicists who by then had repeatedly voiced their disbelief in causality on patently inadequate philosophical grounds.

This hardly flattering story, available in a painstakingly documented book-length monograph,[9] would not have proved culturally effective even if its author had seen what was really wrong with Heisenberg's claim.[10] For then as now physicists were unmindful of the fact that one has to rely on metaphysical considerations to conclude that sequences of empirically observed events, that do not depend on man's own immediate actions, embody causal interconnectedness. Also, for purely logical reasons, the recognition of causality cannot depend on whether a particular interaction can be measured exactly or not. Insensitivity to this purely logical point was the reason why physicists paid no attention when in late December 1930 they read a letter in *Nature*, written to its editor by a prominent British philosopher, J. E. Turner, of the University of Liverpool. In that letter Turner took to task a leading British physicist, the future Nobel-Laureate G. P. Thomson, who had just claimed that "physics is moving away from the rigid determinism of the older materialism into something vaguely approaching a conception of free will." It was not difficult for Turner to notice that Thomson based his doubts about causality on equivocations with words. Or to quote Turner: "Every argument that, since some change cannot be 'determined' in the sense of 'ascertained' it is therefore not 'determined' in the absolutely different sense of 'caused', is a fallacy of equivocation."[11]

This fallacy has become the very dubious backbone of all claims that epistemology is to be drastically reformulated in terms of quantum mechanics, including its latest refinements in terms of Bell's theorem. Underlying all those claims is the contention that physical interactions are non-causal because they cannot be localized by exact measurements. This contention has nothing more for its basis than the wholly gratuitous argument that "an interaction that cannot be measured exactly, cannot take place exactly."[12] In this phrasing, which is perhaps more easy to follow than the one by Turner, one is faced with an equivocation, namely, with taking the same word "exactly" in two very different senses. The inability of measuring exactly is a purely operational failure, which depends on the tools, conceptual and instrumental, available for physics at a given time. The inability of an interaction to take place exactly is an ontological defect which is totally independent of one's ability to measure exactly its quantitative parameters. To infer from the operational inability to an ontological defect is a jump in reasoning which the Greeks of old had already branded with a special name, *metabasis eis allo genos*. On that logical jump, or rather fallacious step, rests that epistemological epidemic which modern physics has, in terms of quantum mechanics, bequeathed to twentieth-century culture.

The culminating point of that epidemic is the heedless reduction of external reality to one's perception of it, the old philosophical fallacy of *esse est percipi*, which is now wrapped in the ever more esoteric mathematical formalism of the latest developments in quantum mechanics and probability theories. Only a few thriving in that rarified atmosphere would emulate the consistency of one of the foremost spokesmen of quantum mechanics, still alive. In a conversation with me, about eight years ago, he acknowledged that if a thief had taken his wallet, he would not have the right to say that his wallet had been really taken, but only that he had the impression of his wallet having been taken. Courtesy prevented me from remarking that if he had gone to the police station with that claim, he would be the first to be detained, and perhaps his university would be asked to send him into early retirement lest he spread further the disease of not seeing the difference between subjective impressions and objective reality.

An epidemic is doubly dangerous when it is not recognized to be such. Most physicists have in fact gloried in what is nothing short of a deadly intellectual disease. A brief but telling example of this is in Max Born's book *My Life and my Views*, which came out in English in 1963. "I am convinced," he wrote there "that theoretical physics is actually philosophy. It has revolutionized fundamental concepts, e. g. about space and time (relativity), about causality (quantum theory), and about substance and matter (atomistics). It has taught us new methods of thinking (complementarity), which are applicable far beyond physics."[13] In saying this Born condensed into a few lines what has become a modern philosophical and cultural creed. Worse, unless some far more reputable Creeds that have been the object of endless scrutinies, this kind of creed has been accepted with no further ado.

Undoubtedly, physics is far more philosophical, and indeed metaphysical, than most physicists would dare to think. Also, the truth of any philosophy demands much more than its being voiced by many, or almost all, in a particular age. Further no philosophy can dispense of a definition of the terms of its basic claims. One would look in vain in Born's book for a definition of what he really meant in speaking of the "revolutionizing of fundamental concepts." Are the terms "revolutionize" and "fundamental" really understood or clarified just because they are used day in and day out? Again, does a physicist cut a trustworthy philosophical figure when, faced with questions about the reality of time, he resorts to the lame excuse that he was talking about time merely as a physicist? But this is precisely what Einstein did in 1922, and at the Sorbonne at that, where Bergson exposed the weakness of Einstein's claim that relativity has fundamentally changed man's understanding of time.

Concerning Born's remark about the revolutionary reshaping of our thinking about causality, the foregoing remarks should suffice. Neither in his writings nor in the writings of other modern physicists have I found a single reliable summary of the origin and development of the philosophical doctrine about substance. Had they studied Aristotle, they would have found that his doctrine of substance cannot be revolutionized by any physics. The reason for this is simple. In speaking about substance Aristotle postulated a reality which by definition was unobservable.

In Aristotle's *Metaphysics* twentieth-century physicists could also have found the vitally serious reason for which Aristotle made that postulate, a reason for which modern physicists do not seem to care at all. It lies in Aristotle's meditation on the epistemological fiasco produced by the contradictory contentions of Heraclitus and Parmenides. For the former everything was change, for the latter all change was illusory. Aristotle postulated an unobservable entity which stands firm under any and all change, an entity aptly called *sub-stance*. He perceived that only through that postulate can one logically maintain a connection between the starting point and end point of any change and thereby assure coherence to any discourse in this world of change. For rationality stands either for coherence or it is not rationality at all. As they claim to revolutionize the notion of substance, modern physicists should at least be attentive to what is demanded by elementary logic.

While space, time, causality, and substance are notions that almost immediately reveal their very complex nature, the idea of complementarity as a new mode of thinking given us by modern physics seems to be within the easy reach of any and all. What that mode of thinking has actually fostered has been the illusion that it is possible to dispose of ontological reality. The chief culprit in this respect is Niels Bohr himself. I am, of course, talking of Niels Bohr the philosopher and not of Niels Bohr the physicist.

More than one philosopher of science has tried in recent years to present Bohr as an epistemological realist.[14] None of them has paid attention to the conclusion of the still most extensive study on Bohr's epistemology, a study now almost twenty years old. There it was pointed out that in promoting the principle of complementarity Bohr wanted to prove that it was possible to get around ontology.[15] In fact, whatever has been disclosed about the cultural and philosophical roots of complementarity as held by Bohr invariably points to anti-ontological tendencies.[16] It is a measure of Bohr's pseudophilosophical success that ontology is held up to implicit or explicit ridicule in books by prominent physicists. Examples are the claim that questions about being are not profitable[17] and the assertion that concern for being as such is as much a waste of time as the medievals' efforts to count the number of angels that can be put on a pinhead.[18] Faced with such scoffers at ontology, one cannot do better than to ask whether the moon is there only when one looks at it, the question which Einstein addressed, around 1950, to A. Pais, his future scientific biographer,[19] though hardly in the hope that his question would enlighten many.

Indeed, the champions of the doctrine of complementarity failed to understand that it is not applicable at that basic level of any and all human experience where ontology and ethics have the first and last say. The "to be" is not complementary to the "not to be" in the sense in which colors are complementary to one another. While one is free to commit suicide and thereby destroy one's life, one is not free to come back at will from death to life and repeat that process *ad libitum*. The connection between life and death is not the kind of reversible approximation which exists, through the correspondence principle, between classical and modern physics.

The moral field is no less instructive. The morally right and wrong are not complementary in the sense in which major and minor tones are interchangeable at will and pleasure. Or to take politics. It is easy and tempting to juxtapose the pluralism of democratic parties with party dictatorship as complementary forms of political life, as was done by Bohr, who had more than one word of praise for Stalinist Russia. Very recent events have made all too clear the enormous difficulties of abandoning the single-party system for a plurality of parties, although it has always been relatively easy to introduce party-dictatorship.

Again, it would be mere flippancy to juxtapose free will and mechanical constraint as mere complementary aspects of one and the same reality. What reality? If it is free will, it is not iron-clad constraint. If it is materialistic determination, it is not the freedom of the will. Only a culture heedless of the mysterious though very obvious reality of human free will could find solace in the preachment of modern physicists about the restoration of free will through Heisenberg's uncertainty principle. It is in such a culture that Einstein's ambivalent statements on free will are seen to be philosophically justified in Quine's puzzling remark: "Freedom of the will means that *we* are free to *do* as we will; not that our will is free to will as it will, which would be nonsense."[20] A nonsensical dichotomy between a free man and his free will.

And what about virtue versus sin? Are they complementary in the sense of being interchangeable, just as the wave-aspect and particle-aspect of atomic and subatomic units of matter are believed to be? What about an upright conscience? Is it on the same level as a conscience taken for a social convention or a taboo, one being just as good or bad as any other? Are there not even in quantum mechanics taboos, such as the exclusion principle and various quantum numbers that cannot be tampered with without making shambles of all atomic and nuclear physics? What about the political motto that all men are created equal? Is it merely complementary to the widespread social and political practice that some individuals, races, and nations are distinctly more equal than others?

Perhaps in this place and context, it will not be amiss to take a searching look at the abuses to which Roman Catholic theology has been exposed through the introduction into its method of the doctrine of complementarity. Tellingly, this disgrace was brought on Catholic theology by distinguished theologians who did not have as much as an undergraduate training in physics. Is the authority of Peter's successor a mere complement to the forming of majority opinions through the manipulation of Catholic news media? Are the memorable words of the present occupant of Peter's chair, that "the Church did not invent itself,"[21] a mere alternative to the widespread practice of inventing a new church in every nook and cranny of the Catholic Church, to say nothing of other churches? Is Christology a mere jumble of paradigms, none of them more fundamental than the other? Is the theological understanding of the Church a heap of models, all of them up for an arbitrary choice? Are the words of Christ about the rock against which the gates of hell shall not prevail the mere mirror image of any heap of quickly shifting ecclesial pebbles? What query can be set up as complementary to the question: "What does it profit a man if he gains the whole world and in the process loses his very soul?"

Clearly, if the physical doctrine of complementarity is needed to teach man that moral virtues go in pairs, then one tries to cure a sickly culture by feeding it with mere placebos. Or is our culture so insensitive to anything except science that it needs some dubious warning from science to reawaken to things beyond science? This may be more true than we would dare to think. Here it would be tempting to talk about the efforts to turn various branches of humanities, including philosophy and history, into science. But let me refer to a more mundane domain which cuts us to the quick at our pocketbooks.

I mean the world of advertisement. The barons of the Madison Avenues of New York, Tokyo, London, Paris, Frankfurt, Brussels, and Milan are convinced that by spending billions of dollars on advertising the public will buy anything wrapped in science and sex. Here let me recall only two examples relating to science. In 1972

Time magazine promoted its sales with a page that carried under Einstein's thoughtful photograph the words: "Everything is Relative."[22] The last page of the July 16, 1991, issue of *The New York Times* carried the huge picture of a pig with wings, with the script in bold letters under it: "In a courtroom, anything will fly if a scientist testifies to it." Such was the self-promotion of *Forbes*, a magazine aimed at leading businessmen.

Should we then resort to the science of complementarity to learn that moral virtues go in pairs? Can the pseudo-moral wisdom of complementarity equal the Gospel warning about the need to be wise as serpents as well as meek as doves, one of many similar injunctions of balance in the Scriptures? Does not the Book of Proverbs, which is a vast storehouse of such injunctions, antedate modern physics by two to three thousand years, and will it not be around even when the doctrine of complementarity has become a long outdated phase in physical theories? Is not finality the very word that has no place there?[23] What is going to happen to a culture that tries to purchase wisdom through terms that, if taken rigorously, have to do with centimeters, ergs, farads, hertzs, henrys, and quantas, but nothing at all with virtues and wisdom? Is not the search for sound epistemology in misplaced philosophical interpretations of physics equivalent to the claim, so popular in the heady days of the Enlightenment, that "all errors of men are errors made in physics"[24]?

Had scientists given more attention to such and similar questions, one thing at least might have clearly emerged on their mental horizon. Scientific context, insofar as it is strictly scientific, means no more and no less than the quantitative analysis of the quantitative aspect of things and their quantitative applications. This is what all scientists do as they try to bring their fields closer and closer to the exactness of physics. Beyond this strictly defined sense, the meaning of the scientific context is much more philosophical than scientific.

The failure of scientists to see clearly the larger scientific context has been compounded by many philosophers of our times who have tried to be scientific in philosophy. As Pierre Duhem showed conclusively almost exactly a century ago, in science it is possible to assume some basic philosophical terms on a commonsense basis and go on with one's work about quantities and the quantitative properties of things and their interactions.[25] From this it does not, however, follow that the successful handling of quantities makes one a good judge of those basic terms. To think so would be to dispense of the principal task of philosophy, which is to explain what is meant by the understanding (*episteme*) of reality. Unfortunately, most modern philosophers have either satisfied themselves with the futile task of trying to understand understanding itself, which is putting the cart before the horse. Also, it has become the hallmark of philosophical sophistication to speak of propositions about reality as mere forms—logical, psychological, sociological—and simply assume that such forms have something to do with reality, which is something far more.

The result for the understanding of science and of the scientific context is a morass of misunderstanding, which culminates in the claim of Feyerabend that nothing about science can really be understood.[26] Feyerabend has, of course, merely unfolded in full the premises of Popper, Kuhn, Lakatos, to mention only a few names. One may also say that Feyerabend has completed the circle of futility which Popper began to introduce as something that contains all science. Popper did so by failing to recognize that if falsifiability is the hallmark of a scientific proposition, then this general rule can be scientific only if it is falsifiable. Since Popper refuses to admit this, his philosophy of science is not scientific, whatever else it may be. It has certainly become a major cultural trap with its three worlds and a universe about which he claims that it takes on all possible forms of existence through its eternal evolution.[27] The cult of the universe, or unabashed pantheism, is not an insignificant philosophical abuse of modern science, and especially of its latest branch, cosmology.[28]

And now about culture itself. A mere look at the word culture brings one within sight of a conceptual minefield. It would be possible to get rid of that minefield by proposing a definition of culture, a proposal fraught with great perils. One of them was acknowledged by the chairman of a two-day symposium on "The Permanent Limitations of Science," which I attended earlier this year in California. The panel discussion that followed the first lecture made it clear that the symposium was to demonstrate once more a phenomenologically facetious definition of most symposia. I heard that definition in Athens in 1975, at the opening of an international symposium on culture. Since then I have had many occasions to feel inclined to believe that, to quote that definition, "Every symposium begins in confusion and ends in confusion, though on a much higher level." Sensing what was in store for the rest of that symposium in California, I proposed to the chairman that we begin by defining what science is. My suggestion made him almost hit the ceiling. If we define science, he replied, we will have to close the symposium.

Before giving my definition of culture, I would like to say something about that minefield. Conceptually, it consists of a host of definitions. In 1952 two prominent American anthropologists, Alfred L. Kroeber and Clyde Kluckhohn, canvassed the anthropological literature for definitions of culture and collected one-hundred-sixty-four of them. Almost half a century later, the number could easily be increased to three to four hundred. Among the definitions found by Kroeber and Kluckhohn there are some worth recalling. They found culture to be taken for "learned behavior," a most exclusively inclusive definition. In another definition culture was taken for "ideas in the mind,"a definition, which, if taken at face value, would exclude from culture everything outside the mind, including books, paintings, sculptures, architecture, and even symposia. Similarly self-defeating were such apparently learned definitions of culture as "a logical construct," "a statistical fiction," to say nothing of culture defined as "a psychic defense mechanism."[29] On seeing that bewildering jungle of definitions, one cannot help, at least for a moment, see some merit in Göring's remark: "Whenever I hear the word culture I reach for my gun."

Kroeber and Kluckhohn held that culture was an "abstraction from behavior."[30] They failed to note what a cultural minefield they were laying by that definition. On a behavioristic basis, violence must be recognized as an integral part of culture, that is, something which is to be cultivated. Modern scientific cultures have not failed to provide many instances of violent behavior, both on the small and the large scale. Little comfort can be taken from the remark of a prominent physicist that modern science made it possible to wage war with the least amount of casualties. He had in mind the chasing of the Iraqis out of Kuwait. One can, of course, glory in the mere two hundred or so Western casualties as long as one is allowed to shrug off, to quote the words of an American newswoman, the more than one hundred thousand Iraqi dead—including women and children—as a "collateral carnage."

Clearly, one has to avoid behaviorism in all its forms to gain a better, that is, a truly cultured hold on culture. One indeed has to dismiss much of modern philosophy, which is rudely or subtly behavioristic, in order get out of the conceptual and ethical morass that has grown around the amorphous phenomena taken for culture.

It should be enough to think of the legalization of sundry forms of cohabitation on the ground that they are all are mere cultural patterns, none better than the other, or rather equally worth being cultivated. It is the preoccupation of most modern philosophers with patterns— of thought or of behavior—that reveals their conscious or tacit vote for behaviorism. Patterns represent a perpetually shifting ground where no stable place is reserved for asking the question about the ground itself.

Here too some prominent physicists lent a helping hand to encourage philosophical sleights of hand. It is enough to think of the brief paragraph whereby Eddington brought to a close his *Space, Time and Gravitation*, his famous interpretation of general relativity for the broader cultural context. There he wrote: "We have found a strange footprint on the shores of the unknown. We have devised profound theories, one after another, to account for its origin. At last, we have succeeded in reconstructing the creature that made that footprint. And Lo! it is our own."[31] Eddington failed to ask whether the footprint could be real if the shores, or the ground, too, were the creation of his idealist philosophy.

To ask such and similarly elementary, or rather fundamental questions in epistemology, is to ask whether human understanding demands for its survival that the intellectual grasp of reality, of things, be taken for a court of final appeal. Once in that court, one has to realize also that quite a few terms, whereby that grasp is performed, are mutually irreducible. It should be enough to think of the mutual irreducibilities of qualities to quantities, of values to volume and so forth. Perhaps the collapse of institutionalized Soviet Marxist ideology, which claimed to gain qualities by heaping quantities upon quantities, may awaken diehard Hegelians in their comfortable university chairs in the West.

For nothing short of a cultural awakening would be produced once the fact is granted that the term cult, which lies at the very basis of the word culture, is one such irreducible term. Culture I define as the ensemble of procedures, mental and physical as well, strongly or weakly symbolic, that can be described as a cultic action. This definition, in itself descriptive, perhaps even behavioristic, becomes an epistemological definition as soon as one takes the cultic action for what it truly is, a form of worship. Worship is more than reverence or preference. Worship is adoration, a service, a total subjection, a complete surrender to something or somebody else. The proverbially foremost manifestation of such act or attitude is, of course, religion.

Worship or religion is of many kinds, depending on what is worshipped. The objects of worship have a frighteningly wide range: pleasure, health, fame, power, party, society, the self, the universe, and God, to mention a few major ones. Being different from one another, those objects of worship allow an objective discrimination among cults, or religions, and therefore cultures themselves. A culture, as was noted before, is characterized by the cultic action which dominates it, and it is its object or target or supreme goal which defines the nature of such action. Culture is worship of something or somebody.

But there is more to all this than an intimate connection between religion and culture. That connection had an able spokesman in the late Christopher Dawson, although, curiously enough, Dawson did not emphasize the fact that even etymologically culture is rooted in cult.[32] One would, of course, look in vain for that connection in Matthew Arnold's *Culture and Anarchy*, the first major meditation on culture. When it was written in 1869, the word culture had not been widely used in English for more than a generation. As a chief promoter of Anglican modernism, Matthew Arnold had to avoid the word cult as it brings to religion that vivid concreteness which modernists of all sorts take for a sting.

The sting could easily be taken out of culture as religion once culture became defined by Arnold as "the love of perfection; it is a study of perfection."[33] Perfection, which for many Christian centuries had been taken for the fruit of religion, was now seen as the derivative of pursuits with little if any emphasis on religion as understood in the Christian tradition. Arnold's sympathies were with the Hellenizers whom he contrasted with the Hebraizers, embodied for him in the Puritans and their descendants. Beyond that he tried to be as vague as possible. Vagueness, or the art of being inclusively exclusive, has always been a standard means of keeping at bay the issue on hand. A studied exercise in vagueness was the gist of Arnold's next and most often quoted definition of culture as "to know, on all the matters which most concern us, the best that has been said and thought in the world."[34]

Religion, insofar as centered in a personal Creator who also specifically revealed himself and therefore demanded a most specific cult, was to be the great loser in Arnold's celebration of culture. The precipitous dechristianization of the Western World during the last fifty years or so has provided many telling details about this. One of them is a remark of the Viscount of Norwich, a political ally of Winston Churchill, who recalled that his constituents wanted only to be assured that he was not a Roman Catholic and added: "For the majority of English people, there are only two religions: Roman Catholic, which is wrong, and the rest which don't matter."[35] Half a century later this mentality, once the dubious distinction of the upper classes, has taken hold of all classes.

In speaking of Matthew Arnold it is impossible not to think of T. H. Huxley, two memorable opponents in a literary debate about the respective merits of scientific and literary cultures. Publicly, T. H. Huxley tried to be evenhanded: "There are other forms beside physical science, and I should be profoundly sorry to see the fact forgotten, or even to observe a tendency to starve, or cripple, literary or aesthetic culture for the sake of science."[36] He was not evenhanded to the extent of including religion in those other forms. In private he dismissed religion insofar as it is the cult of someone infinitely higher than man, because he cherished a cult which, in the guise of science, allowed the cult of the scientist himself, again in the guise of the cult of his scientific attainments.

For the proof of this, one need only recall what happened following the death of Huxley's only son, at the tender age of seven. Of course, the Reverend Charles Kingsley, Chaplain to the Queen and Huxley's fellow member in the Athenaeum Club, was too much of a modernist to sound convincing as he tried to comfort Huxley with reference to resurrection and eternal life. Huxley found no merit in Kingsley arguments: "It is no use to talk to me of analogies and probabilities. I know what I mean when I say I believe in the law of the inverse squares, and I will not rest my life and my hopes upon weaker convictions. . . . Science seems to me to teach in the highest and strongest manner the great truth which is embodied in the Christian conception of the entire surrender to the will of God."[37]

Such an apotheosis of science, which has many spokesmen today, even some most well-meaning ones,[38] becomes the more self-defeating the more elaborately it is presented. This was true in Huxley's case as well. He did not notice the conceptual or logical trap he opened up for himself. Between the lines Huxley argued that the Christian cult or culture as advocated by Kingsley was not empirical or scientific enough, although it had already made many concessions to a scientific method taken for the only method worth using. Then as now, liberal theologians prided themselves on being empirical. But they did so in such a way as to deprive themselves of their greatest factual evidence.

Only Christian theologians faithful to the greatest fact of all history, the fact that also made Christian culture, can turn the tables on all those, scientists or not, who endorse the most poignant lines in the exhortation which Huxley addressed to Kingsley: "Sit down before facts as a little child, be prepared to give up every preconceived notion, follow humbly wherever and to whatever abysses nature leads, or you shall learn nothing. I have only begun to learn content and peace of mind since I have resolved at all risks to do this."[39] Those theologians only need to ask Huxley and his ilk whether they have ever sat down before the fact of Christ, the only religious prophet who said that one has to be a child in order to get hold of Him.

Huxley's cultic reverence for science can easily be matched with momentous admissions about that cult's failure to live up to its promises. In taking a stock of a long life spent in single-minded pursuit of scientific discoveries, S. Chandrasekhar said: "I have a feeling of disappointment because the hope for contentment and a peaceful outlook on life as a result of pursuing a goal has remained unfulfilled. . . . A [fulfilled] life is not necessarily one in which you pursue certain goals. There must be other things."[40]

In addition to science, the full range of "other things" is needed to have culture. But the fullness of those other things does not belong to one and the same level. Those levels form a hierarchy along the parameter of ontology as well as along the parameter of values. Such a hierarchical ordering may evoke the idea of a cone, which in turn may suggest the idea of a straightforward integration from top to bottom and from bottom to top into a single conceptual system.

Nothing can be more misleading. Even in mathematics integration works only under very specific assumptions. In physics, one has to resort all too often to the method of mere summation. Further, no success with summation or with integration does assure the integration into one of all physical forces. Integration becomes a big order that cannot be delivered when one moves out from the quantitative level. Non-quantitative concepts are at best partially overlapping, making the task of "integration" a linguistic impossibility. This is not to say that mutually irreducible concepts are contradictory. But they remain separate tools of grasping this or that aspect of reality and reality itself.

Even apart from the infinite difference between Creator and creatures, the full range of finite ontological reality cannot be compared to a single cone as if spirit and matter were of the same stuff. Only when one deals with matter alone can one represent reality as a cone to which the truth about conic sections can be applied in a very specific respect. Wherever the cone is cut, it reveals the same quantitative patterns. All such sections—parabolas, hyperbolas, ellipses, circles—are therefore on the same level because they can be transformed into one another through purely quantitative operations.

Herein lies the greatness and poverty of science as a cultural factor, regardless of whether one deals with Newtonian physics or with modern physics. Physics is supremely competent about all material processes because they all obey quantitative patterns. This point was cultically stated long before there was science, Newtonian or modern. It was through attention to the phrase, "He [God] arranged everything according to measure, number, and weight, in the Book of Wisdom (11:20) that trust in the mathematical ordering of nature has become a cultural climate long before there was mathematical physics.[41] With respect to their quantitative aspects, all material things, including the brain processes that sustain thinking and free will, are on the same level. In that sense, science is the great and foremost leveler of all.

Herein lies a great warning to all those who dream about a scientific culture. Science, which can say nothing about that act of cult or worship which makes culture what it is, cannot even give assurance about the very existence of things whose quantitative properties it investigates. To be sure, science shows in its own way that all things are limited insofar as they are measurable. In this way science adds a quantitatively specific touch to that far deeper limitedness of things, including their totality, or the universe,[42] which is their contingency, and reveals their utter dependence on the Creator for their existence. To distill this very important cultural contribution from science it is necessary to do a thorough work in the vineyard of philosophy and theology.

Theologians disdainful of the luminous rationality of the inference from things visible to their Creator have been fond of looking in science for guidance in epistemology. One such theologian was Harnack. Here I would recall two details about him. He cautioned his students against drawing openly from his rationalistic theology the denial of the divinity of Christ and of the resurrection of the body, and in fact a denial of all the supernatural. Contrary to Harnack's counsels, his students, soon to occupy the most prestigious chairs of theology in German universities, quickly made a public sport of such denials, in the cloak of the demythologization of Christianity. The other detail was preserved by no less a physicist than Arnold Sommerfeld who wrote: "I am told that Adolf Harnack once said, in the conference-room of the University of Berlin: 'People complain that our generation has no philosophers. Quite unjustly: it is merely that today's philosophers sit in another department, their names are Planck and Einstein'."[43]

I have to disagree thoroughly, and not so much because good philosophers have become almost scarcer than hen's teeth. My main reason concerns the role which prominent physicists have played in spreading the idea that uncaused causes are operating everywhere in the material world at its most fundamental levels. If such claims, which have become almost *de rigueur* in scientists' talk about culture and epistemology,[44] are true, culture is finished as a coherent texture. Then everything happens, which means that nothing happens, to recast in modern form the epistemological and ontological dead-end formulated by Heraclitus and Parmenides. Those physicists who claim that they can produce that everything which is the universe, are still to produce through their scientific skullduggery much smaller things. Such a smaller thing would be a culture, which, incidentally is the indirect chief target of Darwinistic ideology as distinct from evolutionary science. No less a Darwinist of our times than G. G. Simpson of Yale University identified the chief achievement of Darwinism as the abolition of substances.[45]

At any rate, if nature is sustained by the operation of uncaused causes, everything becomes possible and lawful, however lawless. For in that case there will be no things and no laws. In that case, whatever there remains of culture will be dominated by the mush-rooming of weird subcultures, many of them already occupying the center stage of publicity. Against that development mainstream cults or religions would protest in vain, if it is true that nature is run by uncaused causes. In fact even the word mainstream then becomes a contradiction in terms.

All this is a matter of plain logic. It is a symptom of the cultural sickness of the West that this logic was better kept in mind in the very political system which boasted of being scientific and whose sudden collapse constitutes the major cultural event of our century. The witness is the late Arthur Koestler, an erstwhile admirer of Communism. In the most frightening pages of Koestler's *Darkness at Noon*, one reads of the brainwashing of Rubashov who has to listen to a commissar's sinister suggestion: "The temptations of God were always more dangerous for mankind than those of Satan. As long as chaos dominates the world, God is an anachronism."[46]

So much for those who see redeeming cultural value in the science of uncaused causes and in its latest offshoot, the physics of chaos. Once more one witnesses a philosophical rape of good physics. It is also a rape of that good sense which kept the expression "uncaused Cause," however inadequate, for the One whose worship has been the pledge of a culture where nature, man, and his work, be it science, do not readily become objects of cultic reverence. The Christian matrix of the origin of Newtonian science is a subtle proof[47] that Christ was true, even in respect to science, to his promise: "Seek first the Kingdom of God and the rest will be given to you." The rest included science and culture and the understanding of both in terms of true cult.

[1] Quoted in G. Holton and Y. Elkana, (eds.), *Albert Einstein: Historical and Cultural Perspectives: The Centennial Symposium in Jerusalem* (Princeton: Princeton University Press, 1982), p. xv.

[2] For details, see ch. 1 in my *The Absolute beneath the Relative and Other Essays* (Washington, D. C.: University Press of America, 1988).

[3] A. S. Eddington, *Space, Time and Relativity: an Outline of General Relativity* (Cambridge: University Press, 1920), p. 82.

[4] See M. Planck, *Wege zur physikalischen Erkenntnis: Rede und Vorträge* (4th ed.; Leipzig: S. Hirzel, 1944).

[5] W. De Sitter, *Kosmos* (Cambridge: Harvard University Press, 1932), p. 108.

[6] Some of that literature is quoted in my essay, quoted in note 2 above.

[7] For details, see ch. 11, "The Quantum of Science," in my Gifford Lectures, *The Road of Science and the Ways to God* (Edinburgh: Scottish Academic Press; Chicago: University of Chicago Press, 1978). Italian translation, *La strada della scienza e le vie verso Dio* (Milano: Jaca Book, 1988).

[8] "Ueber den anschaulichen Inhalt der quantentheoretischen Kinematik und Mechanik," *Zeitschrift für Physik* 43 (1927), p. 197.

[9] P. Forman, "Weimar Culture, Causality, and Quantum Theory 1918-1927," in *Historical Studies in Physical Science* 3 (1971), pp. 1-106.

[10] For an analysis of Heisenberg's famous essay of 1927, see my essay, "Determinism and Reality," in *Great Ideas Today 1990* (Chicago: Encyclopaedia Britannica, 1990), pp. 277-302.

[11] *Nature,* Dec. 27, 1930, p. 995. Turner referred to Thomson's book, *The Atom* (London: T. Butterworth, 1930), p. 190.

[12] This is my reformulation of Turner's phrase. I first used it in 1972 in an article which later formed the first chapter of my book, *Chance or Reality and Other Essays* (Washington: University Press of America, 1986).

[13] Max Born, *My Life and Views* (New York: Scribner, 1968), p. 48.

[14] See for instance, ch. 1 in H. Krips, *The Metaphysics of Quantum Theory* (Oxford: Clarendon Press, 1978). Such efforts find insoluble difficulties in Bohr's flat declaration that "there is no quantum world. There is only an abstract quantum physical description. It is wrong to think that the task of physics is to find out how nature is. Physics concerns what we can say about nature." See A. P. French and P. J. Kennedy (eds.), *Niels Bohr. A Centennial Volume* (Cambridge, Mass.: Harvard University Press, 1958), p. 305.

[15] C. A. Hooker, "The Nature of Quantum Mechanical Reality: Einstein versus," in R. C. Colodny (ed.), *Paradigms and Paradoxes: The Philosophical Challenge of the Quantum Domain* (Pittsburgh: University of Pittsburgh Press, 1972), pp. 67-302.

[16] See G. Holton, "The Roots of Complementarity," *Daedalus* 99 (1970), pp. 1015-1055.

[17] R. Schlegel, *Completeness in Science* (New York: Appleton-Century-Crofts, 1967), p. 146.

[18] Such was Pauli's slighting of Einstein's concern for reality (which the latter equated with the possibility of measuring things exactly), in his letter of April 15, 1954, to Born, who was greatly upset by his inability to bring Einstein around to the Copenhagen interpretation of quantum mechanics. Pauli referred to O. Stern as the one who had already used that simile. See *The Born-Einstein Letters* (New York: Walker and Company, 1971), p. 223.

[19] See A. Pais, *'Subtle is the Lord . . .': The Science and the Life of Albert Einstein* (Oxford: Clarendon Press, 1982), p. 5. Pais first reported this detail in his essay, "Einstein and the Quantum Theory," *Reviews of Modern Physics* 51 (1979), p. 907.

[20] Quine made this statement in an interview on the BBC with Bryan Magee, *Men of Ideas: Some Creators of Contemporary Philosophy* (London: British Broadcasting Corporation, 1978). p. 43.

[21] This most felicitous statement of the Pope was made during his visit in the Netherlands where such experiments were particularly in vogue.

[22] September 24, 1979, page facing p. 64.

[23] This is not to suggest that physicists may not stumble on the true structure of the physical world. But even when they do so, they cannot be sure that what they have found is the last word. The reason for this lies in Gödel's theorems. For details, see ch. 4, "Gödel's Shadow," in my *God and the Cosmologists* (Edinburgh: Scottish Academic Press, 1989). Italian translation, by L. Gozzi, *Dio e i cosmologi* (Città del Vaticano: Libreria Editrice Vaticana, 1991). For further details, see my essay, "The Last Word in Physics," *Philosophy in Science* 5 (1993), pp. 9-32.

[24] [Baron d'Holbach], *Système de la nature* (London: 1775), p. 19. Condorcet said the same in his plan for the reorganization of French education, written on behalf of the revolutionary government as well as in his far better known *Sketch for a Historical Picture of the Progress of the Human Mind*, tr. S. Hampshire (New York: Noonday Press, 1955), p. 163.

[25] See his essay, "Physique et métaphysique" (1893), reprinted in *Pierre Duhem: Prémices philosophiques*, edited with an introduction by S. L. Jaki (Leiden: E. J. Brill 1987) and, of course, in his masterpiece, *La théorie physique: son objet et sa structure,* first published in 1906.

[26] This claim of P. Feyerabend's culminated in his *Against Method* (London: New Left Books, 1975), which quickly became translated into Spanish, Italian, Portuguese, Swedish, Dutch, German, French, and Japanese, in proof of the craving which modern culture has for its own disintegration and of its illusion that science justifies that craving.

[27] In this respect Popper merely rehashed ideas dear to Bergson, Morgan, and Whitehead, to mention only some prominent names. For details, see my *The Purpose of It All* (Washington: Regnery Gateway; Edinburgh: Scottish Academic Press, 1990), pp. 134-140.

[28] For details, see my *God and the Cosmologists*, quoted in note 23 above, pp. 58-60.

[29] Such is a selection of definitions of culture in "The Concept and Components of Culture" in *Encyclopedia Britannica: Macropedia,* vol. 16, p. 874, with a reference to A. L. Kroeber and C. Kluckhohn *Culture: A Critical Review of Concepts and Definitions* (Cambridge, Mass.: Peabody Museum, 1952). It is not pointed out either in that article or in the vast essay of Kroeber and Kluckhohn that culture has much to do with cult. Among the definitions quoted in that essay, I find most sophisticatedly misleading the one given by Carver in 1935: "Culture is the dissipation of surplus human energy in the exuberant exercise of higher human faculties" (p. 52).

[30] Kroeber and Kluckhohn, *Culture*, p. 89.

[31] Eddington, *Space, Time and Gravitation*, p. 201.

[32] Thus the word 'cult' does not occur in the index of Dawson's *The Historic Reality of Christian Culture* (New York: Harper Torchbooks, 1960).

[33] M. Arnold, *Culture and Anarchy: An Essay in Political and Social Criticism* (1869; London: Smith, Elder & Co., 1891), p. 6.

[34] Ibid., p. viii.

[35] *Old Men Forget: The Autobiography of Duff Cooper (Viscount Norwich)* (New York, E. P. Dutton, 1954), p. 128.

[36] "Scientific Education: Notes of an After-Dinner Speech" (1969), in T. H. Huxley, *Science and Education* (New York: Philosophical Library, 1964), p. 110. On the debate between Huxley and Arnold, see my essay, "A Hundred Years of Two Cultures," in my *Chance or Reality and Other Essays* (Washington D. C.: University Press of America, 1986).

[37] Huxley's reply was written on September 23, 1860. For its text and that of Kingsley's letter, see L. Huxley, *The Life and Letters of Thomas Henry Huxley* (London: Macmillan, 1900), vol. 1, p. 218-19.

[38] See the conclusion of M. Polanyi's *Personal Knowledge*: Towards a Post-Critical Philosophy (1958; Harper Torchbooks, 1962), where, in portraying the scientific man's endless penetration into the unknown, he states: "And that is also, I believe, how a Christian is placed when worshipping God" (p. 405).

[39] See note 39 above.

[40] K. C. Wall, *Chandra: A Biography of S. Chandrasekhar* (Chicago: University of Chicago Press, 1991), p. 305.

[41] The foregoing passage from the Book of Wisdom was the most often quoted biblical passage during the Middle Ages according to E. Curtius, a foremost student of medieval literature. See his *European Literature and the Latin Middle Ages*, tr. W. R. Trask (London: Routledge and Kegan Paul, 1953), p. 504.

[42] The topic, already discussed at some length in my *God and the Cosmologists*, is fully treated in my Forwood Lectures, *Is There a Universe?* (Liverpool University Press, 1993).

[43] A. Sommerfeld, "To Albert Einstein's Seventieth Birthday," in P. A. Schilpp (ed.), *Albert Einstein. Philosopher-Scientist* (1949; Harper Torchbooks, 1959), vol. 1, p. 97.

[44] Polanyi merely offered a compounded confusion of ontology, field theory, and mathematical probability in writing in his *The Tacit Dimension* (Garden City: Doubleday, 1966, pp. 88-89) that "quantum mechanics has also established the conception of uncaused causes, subject only to the control by a field of probabilities."

[45] Other leading Darwinian gurus said the same. Thus J. S. Gould, in his review of *Simple Curiosity: Letters from George Gaylord Simpson to his Family, 1921-1970*, ed. L. F. Laporte (Berkeley: University of California Press, 1988), wrote that Simpson was fully committed to the Darwinian view that variety is all and essence is an illusion (*The New York Times Book Review*, Feb. 14, 1988, p. 15).

[46] A. Koestler, *Darkness at Noon* (1941; New York: New American Library, 1961), p. 134.

[47] See especially ch. 2. in my *The Savior of Science* (Washington, D. C.: Regnery Gateway, 1988; Edinburgh: Scottish Academic Press, 1988). Italian translation, *Il Salvatore della Scienza* (Città del Vaticano: Libreria Editrice Vaticana, 1992). A Hungarian translation was published in 1990, a Russian in 1992, a Polish in 1994.

14

The Paradox of Change

Dear Graduates: You are now together for the last time as students of Drew University. I am sure you will be back often for reunions. This is my first prediction. I will make further predictions for two reasons. One is that Commencement speakers are expected to know something about the future, especially of the future which is in store for the graduates. My other reason relates to my topic, the paradox of change. Change means future. There is no future unless there is change.

I am rather reluctant to predict that this campus will see the replay of what actually happened at Seton Hall, my university. Some thirty years ago a young man who had graduated five years earlier showed up for reunion. He did not come alone. He brought along a wife and a child of two. He took them to the dorm where he had a great time for four years and introduced his wife and child to the resident priest there. Then the young man and the priest started talking about the good old days, when the coke bottle was larger, the pretzel crunchier,

Baccalaureate address, Drew University, May 22, 1998

and, imagine, even some of the bull sessions were on some literary topics, and not merely on baseball. You know, it is not only the future, but also the past that can be largely a matter of imagination.

The conversation went smoothly, indeed so smoothly that the priest decided to bring up a rather touchy matter. He remembered that whenever the father of the young man had come to visit, an angry confrontation developed between the two. So the priest asked the young man: how are things nowadays between you and your father? The young man beamed. Nothing could be better, he answered. The priest did not want to believe his ears. How come? Well, the young man continued: My ol' man has learned a thing or two during the last five years.

Here is a classic case of a learning experience. I hope that this is not the kind of learning experience which is in store for any of you. But you will learn a great deal. This is my second prediction. And I feel very certain of this.

And now my third prediction, which I present as the most general, the most universally valid, and the most profound of all predictions ever: It is simply this: You will see many changes, indeed a great many more than you can dream of.

When I entered elementary school in Hungary in 1930, in the Late Middle Ages, we learned to write with a narrow piece of chalk on a slate tablet. It was not until the second grade that I had a little notebook. When we advanced to spiral notebooks, we felt as if we had been promoted to younger faculty. Loose leaf binders we looked at as thirty years later people looked at the first landing on the moon.

Forty years ago it was not yet customary for students to come with a tape recorder to classes. The reason is simple. A portable tape recorder operates with semiconductors. In the 1950s tape recorders were still bulky things, mainly because they operated with electron tubes. In 1954, in my first year of graduate school in physics, there was still a full semester course on electron tubes. By the early sixties they had been replaced by semiconductors that opened changes in technology that nobody had thought possible.

The replacement of electron tubes by semiconductors reminds me of a futurologist, that is, the kind of historian who always drives in one-way streets in the wrong direction but never gets a ticket. You see there is a difference between real life and academic life. In 1948 our futurologist was very much concerned about the enormous difficulties of space travel. By then much of the war-time secrecy about the ENIAC and other first generation computers had been lifted. There were photos showing those huge calculating machines filling large rooms, partly because they had refrigerators attached to them. The tens of thousands of electron tubes inside those computers generated considerable heat and this had to be kept under control. This weighed heavily on our futurologist who came to the conclusion that space travel, which needed computers, was hardly possible. The spaceships could not afford extra technicians, whose sole business would have been to replace electron tubes that regularly break down.

At this point you will be tempted to bet that I shall go on to talk about traveling to Mars, then to Europa, one of Jupiter's moons and even beyond the outer limits of the solar system. Let me stay for a while with computers as agents of change. The first PCs came to the market in 1984. Most of you, dear graduates, were at that time in second grade.

At that time all PCs were still unwieldy slow monsters compared with what you can get today for one third of the price. Those PCs were especially reluctant to do parallel columns. Almost each time there was a crisis with the operating system. Now new classrooms come with desks, each with a jack for your PC notebooks. Today the average notebook-computer is ten times more powerful than the computers that helped man land on the moon.

I confidently make the prediction that your children will come to Drew University with notebook computers that will have built-in voice-recognition software in them. This means that your children will not have to take notes during classes. They can safely doze off as the teachers speak because their computers will transform what they hear into a text that right away will appear on the screen of their notebooks. Going to college will be sheer pleasure. It will be nothing but fun. Of course, the exams will remain a problem, but who knows. Some politicians will have an answer even to this.

On a much vaster scale, computers have made possible the greatest geopolitical change of this century. Perhaps I can force your memory to go back to February 1991, when you were sophomores in high school. It was the time of the Desert War, which made it public knowledge that militarily the Soviet Union was no match for the United States. Within one year there was no Soviet Union. The Cold War was over, with its enormous tensions and economic dislocations. As for the Desert War, it lasted only four days, and, at least on the American side, it meant no more than 200 casualties.

How did this happen? Well, it was all technology based on semiconductors. Of the long list of technological marvels which includes the stealth bombers, laser-guided missiles, infrared gunsights and so forth, let me say something of the Patriot missiles. They were operated and guided by a satellite in synchronous orbit. The satellite took note of the launching of an Iraqi missile, transmitted the data to the huge computer in the Pentagon, which analyzed the data, returned them to the satellite, which in turn sent certain instructions to the Patriot missile and triggered its launching so that it might intercept the Scud missile still in mid-air. All that took 3 minutes in 1991. Four years ago, somebody in the know told me that at that time the same operation would take only 1 minute. I would not be surprised if I were told that today 20 seconds would be enough to do the same job.

Now the really surprising thing in all this is that it was all taken for a foregone conclusion already in 1976, that is fifteen years before the Desert War. In that year, our State Department sent a delegation to Moscow. There was a big reception in the Kremlin. Questions of war and peace came up as vodka and caviar were passed around. One of the American delegates found himself face to face with Marshall Ogarkov, the Soviet Chief of Staff and asked him point blank: Since the Soviet Union claims that its military strategy is purely defensive, why does it keep four million soldiers under arms? The Marshall smiled. The question, he said, is irrelevant. Why? the American delegate shot back. Because, the Marshall went on, the next war has already been decided. What do you mean, the American delegate asked with astonishment. Because, the Marshall added, war is no longer the matter of soldiers, but a matter of speed, and speed comes with computers. In America, not only the industry but also the colleges are being computerized. In the Soviet Union, even most offices of the Department of Defense are without computers.

Today, as you know, the really hot issue in the United States is not Whitewater, Monica-gate, and Chinese money, no matter how hot these issues may be to some, but Bill Gates and Windows 98. No matter how it will be decided, the decision will determine the rate at which changes will come during the next decade or two.

Those changes will be enormous and will come ever more speedily, in fact at a breathtaking pace. Therefore, please be ready to catch your breath. Because it will be up to you which of three classes of people you will fall into. That everybody belongs into one of those three classes is an old truth. Possibly its best description was given about eighty years ago, by Nicholas Murray Butler, president of Columbia University. "The world is made up of three groups of people. The first group, a very small one—who make things happen; a somewhat larger group,—who watch things happen; and the great multitude—who don't know what happens."

I would merely add that they do not know, because they don't care to know. This is why about half of Americans regularly do not vote.

So I beg you, please try to be one of those who make things happen. Perhaps not on a large scale, because not everybody can become another Bill Gates, and this is perhaps for the better. As you know there was a report about two months ago, a false report to be sure, that Bill Gates had died. According to the report when Bill Gates arrived at the pearly gates, St Peter told him: Bill, you have been awful lucky all your life. Your luck has not deserted you even at this moment. You have arrived here during a ten minute window when we have a special. I mean, St Peter went on, that anyone who arrives during this ten minute period, can choose whether he or she goes to heaven or hell.

Well, so the report goes, Bill Gates said: St Peter, I have never been religious all my life, so I do not know the difference between heaven and hell. Could I see both first so that I could make my choice. St Peter was in a good mood, so he agreed. After spending half an hour in heaven Bill Gates came back to St Peter's office with a very bored look on his face: Always the same chants, always the same bowing. May I see what is in hell? St Peter let him enter hell. There Bill Gates saw beautiful lawns, swimming pools, ice cream stands and bands playing. He sent word back to St Peter that he preferred to stay in hell. The gates of hell shut tight behind Bill Gates.

But this is not the end of the story. Thirty days later St Peter walked by the gates of hell and let those gates spring wide open so that he might see whether everything was in order there. Well just inside the gates there was a huge cauldron with boiling oil in it. Out of the cauldron the head of Bill Gates stuck out, who immediately recognized St Peter. St Peter, St Peter, he started shouting. What is it Bill? Well, this is not the hell you showed me thirty days ago. You cheated me. Not at all, St Peter replied. How come? Bill Gates shot back. I did not cheat you at all, St Peter said. I merely showed you the demo first as this is customary in advanced business practices.

Now many of the changes you will have to face will be just gimmicks with demos. There is nothing new in this. As the French said long ago: plus ça change plus ça reste la même chose, The more it changes the more it remains the same thing. Those who do not care to know what happened will have only themselves to blame. Among those few who make things happen, who bring about changes, there will be some who will be spin masters. They will market their demos. It is up to those who watch things happen to see the difference between the demos and the genuine items.

To see that difference, you will have to sharpen your eyes about the reality of change and its meaning. Whether you like it or not, you will have a philosophical problem cut out for you. Indeed the whole history of philosophy, and by this I also mean the history of the philosophy of science, can be divided into two classes. In one, you will find those philosophers who take the problem of change seriously. In fact, they hold change to be the primary problem for philosophers. In the other class belong those philosophers who pretend that the problem can be ignored. They are not much better than the ones who fail to recognize that the only way to avoid philosophy, indeed metaphysics, is to say nothing.

In fact no word is more loaded with metaphysics than the word *nothing*. When you say it, you do something with your nervous system, yet your mind assures you that you mean nothing and that you are not under an illusion. This is precisely what a computer cannot do, and will never do. If you are ever confronted with any of those brave guys or gals who tell you that you are an artificial intelligence machine, just ask in return: how do you program the *nothing*?

Now that I have said that the word *nothing* is the most philosophical of all words, I have to correct myself. There is one other word, which is even more philosophical. It is the most trivial looking word of all words, in all languages. It is the verb, to be and its variants, I am, you are, you will be, you were and so forth. We all are, we all have been, we all shall be. We live as we change, yet we keep our personal identity, which is our greatest treasure and privilege. Herein lies the supreme form of the paradox of change. It is the paradox of essence.

For changes make sense, or meaning, only if something remains identical during the change, only if there is an essence, a word that comes from the Latin essential, which in turn is a variant on the Latin verb *esse*, or to be. If there is no essence, that is, something that remains the same in all change, it become impossible to make a judgment on any object, on any process, on any change.

Please, believe me, modern physics did not change this situation. When it replaced its own strange construct, the atoms, those mythical absolutely hard balls, it merely put in their place, the protons and the neutrons, and the electrons. Modern physics still cannot divide the electron, but it has divided the protons and the neutrons into quarks. But in order to understand quarks, physicists have to figure out the constituent or essential parts of a quark, which in turn cannot be done scientifically unless the quark is divided into further parts, and so forth ad infinitum. Or as Jonathan Swift said it long ago about naturalists:

... a flea has smaller fleas that on him prey,
And these have smaller still to bite 'em.
And so proceed ad infinitum.

This is why particle physicists never work themselves out of a job. They can only work themselves out of research money. And this they have largely achieved. So please look for employment elsewhere. Biology shall not work itself out of fleas, but it may also find research money running dry. And this bodes ill for the long range prospects of investment. However, on the short run, your prospects for employment look essentially good. But this is merely a mental judgment and not an inside information from Wall Street. There, only the kind of essences are traded that are bound to intoxicate their consumers.

All our mental judgments are openly or furtively about essences. Now this is the kind of judgment, the basic operation of human intelligence, that controls all its other operations. And this is the kind of operation which a computer, an artificial intelligence machine cannot do and shall never perform. Here is my final prediction, which I make with absolute assurance and also with some great concern. Not so much for myself, because I am an old man, but with concern for younger people like you, who grew up with computers, these unparalleled instruments of change.

You will be challenged not only to make sense of any change, but also to figure out for what purpose to use any and all change. And you will have to do all this freely, that is, with full moral responsibility. You will have to do all this in a culture, which is rapidly shedding the last shreds of its cultural heritage, insofar as culture is rooted in cult or religion. And to make this even worse, our so-called culture is boasting about this cultural suicide.

About two months ago, an essay appeared on the op-ed pages of *The New York Times* whose author presented computers, or artificial intelligence, as the last step in putting man in his place, by denying him any special place. First, he wrote, came, the Copernican revolution that removed the earth, and man on it, from the center of the universe and put the sun in its place. Then came the removal of the sun from the center of the galaxy, and after that the removal of our galaxy from the center of the universe. Then came Darwin, who denied man a special place among the living. Then the discovery, so our guru claimed, of life elsewhere in the universe, and the first steps of communication with extraterrestrials. You see, it is easy to take fiction for facts. Now, he went on, computers abolished the difference between mind and matter. And he called this last step the final liberation of man from the shackles of essentialism.

For the next two weeks *The New York Times* (by the way I am not a paid subscription agent for *The New York Times*) published some thirty letters on the subject. Most of the letters were critical, but none of the criticism touched on the essential difference between human intelligence and artificial intelligence. It is the difference, first, between the conditions of knowledge and knowledge itself, and then the difference between the known thing and the fact that one knows it and one knows that one knows this.

In all the enormous changes that are in store for you, please hang on dearly to your intellect. You will have to hang onto it in a growing wasteland called modern culture, that has no other cult except the cultivation of the individuals' whims and fancy so that the Ten Commandments may be changed into Ten Counsels that are elective but not obligatory.

As graduates of this University with a Methodist tradition, try to be very methodical about hanging on to your intellect. For unless an individual wants to thrive on demos, intellectual, moral and technological demos, he or she will have to be methodical in learning the art of distinguishing the demo from the genuine thing.

In trying to live up to this challenge, you will need real cult, real religion. Not merely phrases about religion, not statistics about it, not phenomenological descriptions of it. All these give you only the art of looking at religion as not so innocent, detached bystanders, who judge others, but never take a measure of themselves. None of those clever tricks will give you an inkling of the essence of religion, which is the power to face up to the challenge of to be, to exist, and do this in the welter of change.

Now you will not find my concluding statement strange at all. If there is anything that recommends the religion of which the Bible is the written document, it is the fact that in the Bible God is given a name which is unlike any name ever given to God in any other religion. That name is *I am who is*, or Yahweh.

That God is a father can be found in many other religions. The same is true about other biblical names of God, such as that he is a Lord, a Master, a Maker, and so forth. But only in the Bible will you find that his name is *He who is*. Moreover it is not Moses who makes this statement. He most emphatically states that it was God himself who revealed his name to him. The very fact that this name can be found only in the Bible and nowhere else should suggest that the Bible is much more than a cultural document. It is a depository of a revelation properly so called.

As graduates of this University, you have to carry this message in a methodical way. And you will find, as John Wesley discovered for himself an already very old truth, that the whole world will be your parish. No matter where you land in this increasingly global parish which is our world, you will never run out of opportunity to remind any and all of the paradox of change and the only solution to it.

The solution to the paradox of change is not what Chesterton once said, that the purpose of paradox is to awaken the mind. This is merely the introduction to the solution. The solution to the paradox of change is that unchanging being of whom Psalm 102 says:

Long ago you founded the earth
And the heavens are the work of your hands.
They will perish, but you will remain.
They will all wear out like a garment,
you will change them like clothes that are changed
But you neither change, nor have an end.

These incomparable words are a reflection in the same Psalm on major upheavals, breakdowns in human lives:

He has broken my strength in mid-course,
He has shortened the days of my life.

This in turn is followed by a prayer:

I say to God: Do not take me away
Before my days are complete
you, whose days last from age to age.

May you all, dear graduates, have your lasting security and permanence in your worship of that unchanging God in a world that has more changes in store for you than it had for my generation, or even the generation of your parents. And when you see the greenback losing ever more rapidly its purchasing value, meditate on the not-at-all financial motto printed on all those not-so-almighty greenbacks: The motto is: In God we trust. May He keep all of you in his loving care. With trust in Him the future is yours. Godspeed to all you!

15

Cosmic Rays and Water Spiders

Society makes strange claims on many of us. It expects politicians to be upright, reporters to be truthful, businessmen to be honest, clergymen to be virtuous, and scientists to be men of universal wisdom. Those who have spent much of their life in studying science and religion are often expected to do the impossible: to prove religion from science. It cannot be done, and certainly not when the religion to be proven is left carefully undefined. It never pays to try to firm up a patch of cloud. But to say this is rather risky at a time when "playing church" has become the hallmark of having "mature" religion, though not something authoritatively definite about maturity. No wonder that even the One who claimed to himself all authority under heaven and earth is subordinated to the authority of a "higher criticism" whose champions cannot see higher than themselves.

Originally published in *Spiritual Evolution: Scientists Discuss Their Beliefs,* ed. J. M. Templeton and K. S. Giniger (Philadelphia and London: Templeton Foundation Press, 1998), pp. 67-97. Reprinted with permission.

In speaking about science there is less risk of being trapped in fog-mongering. The reason is simple. On more than one occasion prominent physicists have come up with pithy phrases that go to the very essence of exact science, physics, the very ideal which cultivators of other branches science try to emulate, with more or less success. One such phrase is Hertz's dictum: "Maxwell's theory is Maxwell's system of equations."[1] For those who find this too esoteric, there is Eddington's warning that science, by which he meant physics, "cannot handle even the multiplication table single-handed."[2] Both these phrases should reveal that in any branch of science the amount of fog should be inversely proportional to the mathematics, or quantitative precision, it embodies. It is this precision, quantitative precision, that makes talking about science relatively risk free.

But nothing is so risky or mistaken as to try to perorate on that basis about anything else, including religion. Why then should one, who is trained as a physicist and a theologian, and has studied all his life the relation of science and religion, be expected to prove religion from science? Is it not to ask him to throw caution to the wind? Why expect him to regard religion, usually left unspecified, with a kind of precision which it cannot have even when properly defined?

Thus, even if one leaves, for the moment, religion unspecified, it would still be true that if a religion (or its theology), has no foundation of its own to stand on, it does not deserve to be propped up even by science. Indeed, many years of study and reflection lead me to think that concerning the relation of science and religion there is good reason to begin with a kind of shock treatment, which, for some time, I have grown fond of formulating with a touch of irreverence: "What God has separated no man should join together." By joining, which is man's work, I mean fusing and ultimately confusing. As to the separation, which I think is God's work, I mean simply this: The cleavage which is between the metric and the non-metric, that is, what is measurable and not measurable, cannot be bridged conceptually.

The metric, or measurable properties of anything, are the basic and sole business of science, the very point which is intimated by Hertz's dictum. Religion, in turn, deals with issues that are imponderable also in the sense that they cannot be measured by calipers, etalons, or any of the marvelous devices science has produced in stunning variety. There is no opposition between being a virtuous man and also six feet tall, but of these two properties only one's physical height can be measured. A persons's virtuousness can be put in the scales, but the resulting measure is very different from the one which science can provide. Unless this difference is kept in focus from the very start, while discoursing on the relation of science and religion, all that discourse will turn into an effort to firm up a patch of fog.

This is not to suggest that sharp focusing on this point has been an overriding concern for me ever since I started writing on the subject of science and religion. Origins have a way of hiding themselves even in the light of keen retrospect. But very clear in my memory are two details from my student years that may serve as a starting point in sketching my journey towards that irreverent shock treatment, although they may by themselves indicate a very different direction.

I knew at the age of seven that I wanted to become a priest. I never had a serious doubt on that score, either before or after I was ordained a priest as a member of the Benedictine Order. By the time I was sixteen I knew that I wanted to be a theologian too. My mind, however, was driven in two other directions as well. One was history. Then as now I feel dissatisfied with my grasp of any point as long as its history is hidden from me. Many years later I found that perhaps the most incisive interpreter of exact science in modern times, Pierre Duhem (1861-1916), also felt this way. His monumental work is a proof that such is not a useless concern.

In addition, the kind of curiosity which propels scientific interest has also been very strong with me from early youth. Thus at the age of twelve, I slipped into a lecture hall, where I was the only one below thirty or so, to hear a talk on cosmic rays. I gained the impression that I understood everything. If I grasped anything at all, the credit should go to the lecturer, a Benedictine priest, who for eight years taught me mathematics in the gymnasium (an exacting form of middle school) in my native town, Győr, Hungary. I did not suspect then that twenty years later I would do, in another part of the world, my doctoral research in physics under the mentorship of Dr. Victor F. Hess, the Nobel-laureate discoverer of cosmic rays. Anyone thrilled by cosmic rays, cannot help being intrigued by the cosmos. Not a small part of my work in the history and philosophy of science turned out to be about the stellar universe.

Five years after that lecture on cosmic rays, the same Benedict-ine priest, who also taught apologetics to seniors in the same gymnasium, asked me to read aloud to the class a passage about water spiders from a book whose title I cannot remember. But few things stand out so vividly in my memory as that passage. Almost fifty years later my heart leapt for joy, as if carried back into my distant younger years, when I found in a book, *Biology for Everyman*, by J. Arthur Thomson, the following passage:

> The water-spider (*argyroneta natans*) which spends most of its life under water, makes a tent of silk on the floor of the pool, mooring it to stones and the like by silk threads like tent-ropes. Sometimes the shelter is woven among water-weeds. If the tent is on the floor of the pool it is flat to begin with, but the spider proceeds to buoy it up with air. Helped by a special thread, fixed at the bottom and to water-weeds at the surface, the spider ascends and entangles air in the hairs of the body. Climbing down the rope, like a drop of quick-silver because of the air bubbles, it passes under the silken sheet and presses off the air. The air is caught by the silk sheet, and after many journeys the nest becomes like a dome or diving bell, full of dry air. In this remarkable chamber, dry though under water, the mother-spider lays her eggs, and there her offspring are hatched out. The dry dome may be used as a shelter during the winter, when the spider remains inactive . . . .[3]

The memory of that passage never faded in my mind as the subject of evolution gradually became a consuming interest for me, mainly because failure to understand it properly is devouring so many minds who deserve better. The evolutionary lore has acted as a major roadblock against speaking intelligently about science and religion. This happened not only because agnosticism and materialism have been foisted on Darwinism from its very inception, but also because some tried to cast evolution into a scenario of ever ascending spiritual self-perfection. Long before Teilhard de Chardin came up with his profuse diction about an Omega Point unsupported with sedulous elaborations about the Alpha Point, similar scenarios appeared at regular intervals from almost the moment that saw the several thousand copies of the first edition of *The Origin of Species* snapped up within a few hours.

In view of my emphasis on the radical conceptual difference between the metric and the non-metric, it becomes a foregone conclusion that there can be no such scenario. Such is one of the principal conclusions which gradually became crystallized in my mind during more than fifty years of intense interest in and systematic study of the two fields, science and religion.

Another conclusion may appear positive, but only to those who have a positive appreciation of a philosophy that can stand on its own feet, without begging for handouts from science, and what is worse, from scientists who often know next to nothing about philosophy. This can be the case even when a scientist, like Einstein, has a gut feeling for reality as existing independently of man's thinking about it. When no such gut feeling is on hand, as was the case with Niels Bohr, the outcome is simply pathetic both in itself and in its destructive seductiveness. Witness the conceptual orgy, dressed in the most esoteric scientific technicalities, within which the wave function for the Universe is made to collapse by conscious thought, although no such learned minds have ever tried to do the same with the wave function of a gold bullion, and not even with that of a club sandwich.

To know that hungry pigs will not fatten by dreaming about heaps of acorns, is worth more philosophically than shelves of books whose authors become entangled ever more desperately in the logical fallacies of the starting point of the Copenhagen philosophy of quantum mechanics. One needs philosophy, not science, to see beyond the surface where science unfolds more and more of the enormous degree of specificity of the material universe. Specificity is suchness. It is the registering of suchness that sparks inquiry after causes: why such and not something else? Of course, one need not be a scientist to see suchness everywhere where there is knowledge and inquiry. But the suchness revealed by modern science about every aspect of the living and the non-living is simply astonishing. Suffice to think of the double helix of DNA molecules and of the properties of quarks, which are strictly quantitative, although denoted as flavors and colors. Such specificities, of which libraries can be filled, have a philosophical significance which a theologian can ignore only at his enormous disadvantage in this age of science.

I do not mean the theologian whose God is the precipitate of process theology, adored in Sunday swim-ins, and savored by chewing nuts, cheese, dates, and gulping down glasses of port. True mysticism never goes without a high-degree of self-imposed deprivation. In the absence of this one can at best *talk* of mysticism and even of religion, in disregard of the injunction that one must be a doer of the word of God and not merely a listener to it, let alone a disinterested listener. By theologian I mean the one, learned or not, whose God is the Father in heaven who in turn has to be begged for the daily bread, for the forgiveness of one's sins, for escape from temptations, because he is that only Father who is truly Almighty. The term "Almighty" means, however, that he is the Maker of heaven and earth, or of that All which is the Universe, writ large. The Son logically belongs here as the "Savior of Science," the very title of a book of mine that came fairly late, but perhaps not too late, in my meditations on science and religion. There is nothing wrong with that as long as evolution means maturation and not the pulling of a rabbit out of a hat by, say, giant mutations. Such giants, no more real than the seven dwarfs, belong neither to the genesis of theology nor to the science of genetics.

In fact the Son should be brought in now, however briefly, because the Father created everything in his Word, the Logos. Such is the reason why the Universe has to be fully logical. Were historians of science, so eager to pinpoint trivial starting points, appreciative of the fact that not the Greeks, but Saint Athanasius, so terribly otherworldly for them, formulated in his struggle against the Arians the full rationality of the material universe, a good deal of the alleged opposition between science and religion would vanish like the morning fog. Of course, some theologians too, busy with the relation of science and religion, should do some soul searching. I mean those who inherited a dislike for the Athanasian, or rather Nicene dogma of the Word's consubstantiality just because the word consubstantial cannot be found in the Bible. Yet the Bible certainly teaches that the Father created everything in the Son. Such a Son has to be strictly divine, because even Almighty God cannot subdelegate his power to create to a mere creature, however exalted

First, therefore, the createdness of the universe. Recognition of this is the very minimum without which there can be no religion that includes prayer to a personal Creator. If the theologian is learned and logical (unfortunately these two qualifications need no longer go together), he must assert that reason may safely infer the existence of the Creator. Science can help him greatly, provided he knows what is being done when one infers the reality of something unseen. The word *inference* is crucial, because all human knowledge which relates not to realities directly experienced on the ground floor, points to levels higher up, to realities grasped by climbing mentally to the second floor and to floors even higher. Knowledge is a seven story mountain, with the number seven symbolizing that perfection which goes with any sound inferential grasp of the existence of God.

So much for the assertion that much of our knowledge (even in science) is an inferential knowledge and is such long before God, the ultimate, becomes the object of one's reasoned inference. If this is not kept in mind and one's feet are not kept on the ground floor, or the first floor where data are gathered for higher inferences, one will be overawed by claims of modern cosmologists that their expertise enables them to create entire universes literally out of nothing. Clearly then one can play not only church but one can play God as well. Such is the acme of the new reformation, achieved by a drastic abuse of science.

Hapless thinking about quantum mechanics merely put the icing on a cake that began to be baked two hundred years ago by perhaps the most consummate chameleon in the history of science. I mean Laplace, who kept conveniently changing in order to remain on the crest of the wave as Monarchy turned into Convention, Convention into Terror, Terror into Directorate, Directorate into Empire, Empire into Restoration. I hope and pray that the report about his repentance on his deathbed is true. There is no more sound reason for believing in God than that he is infinitely merciful. This, however, presupposes that one first dissociates oneself from that modern mankind that takes man's fallen condition for his healthy state so that one may glory shamelessly in one's very glory.

Laplace's famous words, "I do not need that hypothesis," may express either sound science, or something utterly unsound, philosophically that is. Two hundred years ago Laplace sold the idea that our present, exceedingly specific physical world evolved from a nondescript or nebulous primordial cloud, about which he knew only that it was very nebulous indeed. When in 1801 Laplace uttered these words, he did not yet feel the social need to profess belief in God. However, he did not have to be a philosopher to sense that an emphatically nebulous entity never calls for the question about its nebulosity, precisely because it verges on the nondescript. If the supposedly original entity, a cosmic nebula, is nondescript, it automatically parades as that ultimate entity about which no further questions are asked. Such is the pseudo-scientific death knell delivered by pantheism to natural theology.

The implicitness of Laplace's "atheistic" or rather "pantheistic" reasoning was eventually made explicit by Herbert Spencer, who spun philosophical fairy tales about the evolution of the non-homogeneous from the primordially homogeneous. As was noted above, inquiry, curiosity, search for causes are sparked by the registering of suchness or specificities. Now if the primordial state is imagined to be utterly homogeneous (nebulous) then it will not prompt one to raise the question why such and not something else, because what is allegedly absolutely homogeneous has no suchness. It therefore begins to pose as the ultimate in intelligibility and being. Such is, in a nutshell, the logic, of modern scientistic materialism and atheism. There is no opposing it unless one goes to the basics of epistemology. Without doing so one cannot counter the icing on the cosmic cake which is busily baked by some luminaries of quantum cosmology.

What these should say is rather that modern scientific cosmology shows across space and time a universe that cannot be more different from the universe that originated from a primordial nondescript state. Whatever is observed by science about physical reality, whether about its present or remote stage, it is specificity. The philosophically sensitive theologian can and must utilize these results of science to strengthen his inference about the existence of a Creator. But he should be on guard, lest he assume that science will do this for him. Science, which "cannot handle even the multiplication table singlehanded," cannot say a word about causality, not even about things real, nor can it prove that there is a strict totality of things, or a Universe. Science, including Einsteinian cosmology, cannot do this for a very simple reason: the truths of science demand experimental verification. But no scientist, no scientific instrument, can be carried beyond the universe to observe and thereby scientifically verify it. The verification of the universe is an eminently philosophical task.

This task can be achieved, as I argued in my Forwood Lectures given at the University of Liverpool in 1992, available as *Is there a Universe?*[4] Youthful but genuine enthusiasm about cosmic rays had to issue in a resolve to fathom the grounds that justify the strict use of the word universe or cosmos. Three years earlier, in my Farmington Institute (Oxford) Lectures, *God and the Cosmologists,* I had already developed this new form of the cosmological argument.[5] It must, however, have been clearly set forth in earlier works of mine, such as the Gifford Lectures, *The Road of Science and the Ways to God,*[6] or else the citation for the Templeton Prize for 1987 would not have mentioned it as one of the reasons why that Prize was awarded to me.

Another reason was my investigation of the basic framework within which the relation of science and religion can be meaningfully dealt with. The principal aspect of that framework is the unique standing of the category of quantities among all the other categories of thought as listed by Aristotle in the *Categories*. I refer to Aristotle with some trepidation, because of the prejudice that a point made over two millennia ago, especially if made by him, cannot be true. In fact, in a sense, much of the modern philosophical agenda is aimed at showing that qualities, action, passion, substance accidents, etc can be reduced to quantities. Such is a chief contention of the Hegelian right and left. Logical positivists claim much the same when they restrict meaning to statements that can be handled in the manner of mathematics. Logic, as they understand it, is but a form of mathematics. Hence the formalism of symbolic logic, which, incidentally, cannot even prove the reality of a book, filled with symbols, strangely resembling those of mathematics.

Here again it was only in recent years that this difference between quantities and all other categories has taken, in my thinking about religion and science, a central place. I wonder what those engineers of optical instrumentation thought of my paper, read at their vast annual meeting in Orlando, Florida, on April 14, 1996, under the title: "Words: Blocks, amoebas, or patches of fog? Artificial Intelligence and the foundations of fuzzy logic."[7] But unless promoters of artificial intelligence machines consider that difference, they will never understand why their blueprints of thinking machines will forever remain very fuzzy indeed.

This point was explicitly made already in the fifth chapter, "Language, Logic, Logos," added to the second edition (1989) of my *Brain, Mind and Computers*, whose first edition (1969) received the Lecomte du Nouy Prize. I do not know whether this point was among those which the author of *Gödel, Escher, Bach,* who cast there an emotion-laden vote on behalf of artificial intelligence, had in mind in saying there that Jaki had made in that book of his some points worth pondering. For he never went on record by pondering them, even as much as by listing them. Such is the new scholarship: mere reference to a problem, without even naming it, passes for its successful resolution. There was no real reply to *Brain, Mind, and Computers,* although none other than Herbert Feigl voiced the need for it. Apparently I said something very important there that cannot be attacked except with a boomerang.

Professor Feigl, then as later, was one of those very rare academics who never tried to brush off scholarship just because it went together with an ideology diametrically opposite to his. Such was his reaction to the typescript of my first major book on science and religion, *The Relevance of Physics* (1966), which he read for the University of Chicago Press. That book contains passages that anticipate those later observations of mine about the strange limitedness of quantitative notions. The broader meaning of the point is, of course, that science is largely irrelevant about most of what human cogitation is about. This has to be clearly realized if one is to speak meaningfully about the relation of science and religion, instead of serving up indigestible dreams about the two being fused together or standing in radical opposition to one another. They are merely different and will forever remain so, mental acrobatics notwithstanding.

Herein lay another reason for me to find in Duhem a very germane soul and mind. Had Duhem achieved nothing except unearthing through heroic research the medieval (and distinctly Christian) origin of modern science he would have already immortalized his name in the genuine annals of the historiography of science. Duhem was also the foremost among the sane philosophical interpreters of physics, in addition to being a first-rate theoretical physicist. I first came across his towering intellect and noble character when a short biography of his written by his daughter fell into my hands. Well, if not even a sparrow falls to the ground without our heavenly Father willing it, He must have something to do with the moments and places when this or that book falls into one's hands. And just in time, because shortly afterwards I heard one of the chief luminaries among still living historians of science dismiss interest in medieval science as a hobby of Roman Catholics, and especially of priests, such as Duhem. He choked when I told him, in the presence of quite a few, that a chief source on Duhem's life is a book written by his only child, a daughter.

Shortly afterwards (around 1961) I read the second edition of Duhem's *La théorie physique* (1914) available also in English translation as *The Aim and Structure of Physical Theory*. One of the two additions to that second edition is what may be Duhem's finest essay, "The physics of a believer." There Duhem begins with answering the charge of a critic (Abel Rey) of the first edition (1906) that his scientific philosophy is "that of a believer." "Of course"—Duhem writes in reply and rebuttal—"I believe with all my soul in the truth that God has revealed to us and that He has taught us through His Church. I have never concealed my faith, and I hope from the bottom of my heart that He in whom I hold it shall keep me from ever being ashamed of that faith: in this sense it is permissible to say that the physics I profess is the physics of a believer."[8]

As a Roman Catholic philosopher and historian of science, mainly of physics, astronomy, and cosmology, I could have never put it remotely as well as Duhem did. And I could continue with Duhem, who noted right there and then, that it was not in that sense that his critic labeled his physics "the physics of a believer." That critic, for whom it was inconceivable that there could be an epistemology between the extremes of empiricism (positivism) and idealism, charged Duhem with injecting religion into physics, precisely because he proceeded between those extremes, or rather mental abysses. It shows something of the darkness dominating the secular academia around the turn of the century, that realist metaphysics, the epistemological middle road taken by Duhem, could be taken for mysticism and therefore for religion.

In the same way some of my critics, unable to see that middle ground, and just as blind therefore to some salient facts of the history of science, grew fond of dismissing my work as that of a Catholic. This blindness they have inherited as a staple feature of post-Enlightenment Western culture, steeped in contempt for the Middle Ages. In fomenting this contempt, often growing into hatred, the heirs of the Reformation have become strange bedfellows with Voltaire, Diderot, Hume, Gibbon and the rest. For these two camps, though for very different reasons, nothing was so vital as to paint the Middle Ages as black as possible. A dangerous tactic, because it prevents the secular historian from seeing plain, vastly documented facts of scientific history, and prevents the Protestant scholar from seeing the inner logic of "private judgment." I would not mention this had not a prominent Protestant theologian, very busy with science, shared with me his anxiety about that logic. His words, uttered in complete privacy, still ring in my ears: "Protestantism logically leads to naturalism." Not that he said anything new with that.

Contrary to the expectations of that critic of his, Duhem became celebrated among philosophers of science, at least in the sense that they keep borrowing from him heavily, without giving him a thimbleful of credit. There is an awful amount of food for thought in the few lines of a postcard that Lakatos once sent to Feyerabend, stating that all of Popper's philosophy is a rehash of Duhem's. This is, of course, true only in the sense that Popper and others borrowed some of Duhem's method without endorsing with him that ontology (realist metaphysics) which alone assures that the method deals with reality and not merely with the calculations of the physicist and with the ideas of the philosophers of science about those calculations, which all too often they cannot follow.

In referring to that postcard Feyerabend gave the impression that he was fully aware of his indebtedness to Duhem. He might have been, but he gave no inkling of this in his printed works. But such is academic objectivity, whose practitioners can get away with conceptual murders once they occupy a prestigious chair. Equally revealing is a facet in Kuhn's philosophy of science, which the author of a doctoral dissertation on his work pointed out about twenty years ago: The principal Kuhnian theses are all in Duhem, except, of course, that irrationality which destroys Kuhn's ideas on scientific revolutions as so many drastic paradigm changes of the human mind. As a loyal son of the Catholic Church, which is continuity incarnate, Duhem would have never cavorted in radical discontinuities either in religion (theology) or in philosophy, or in science. This is why among other things, of which more shortly, Duhem is an *Uneasy Genius* (1984), the title of my half-a-million-word monograph on him.[9]

I was indeed dismissed time and again as a "Catholic" historian and philosopher of science (and a Jesuit to boot, which, being a Benedictine, I am not). This should be fairly understandable. In this age, when science is corralled by any Tom, Dick and Harry in order to gain respectability for his ideas, few things can be so resented by a Protestant as two claims of mine. One, of which more later, is that creative science has always been connected with a middle-road epistemology, which, of course, is incompatible with the Ockhamism inherited by the Reformers. This is, of course, difficult to understand in this age when some Protestant theologians who divorced their field from philosophy gained high repute, for a while at least. Logic can never be exorcised forever.

The other claim, the medieval origins of modern science, is more easily resented, especially when both believing and nominally believing Protestants hang on to the idea, as if it were a life belt for them in this age of science, that science originated under the spiritual impact of Puritans, and perhaps with Luther and Calvin. These last two were sheer mystery mongers, the former rudely, the latter subtly, in their interpretation of Genesis 1, which even today is a touchstone of truth as to whether one is talking sense or nonsense about the relation of science and religion. Those who find this statement of mine outrageous, may wish to consult my monograph, *Genesis 1 through the Ages* (1992), a history of interpretations of that first chapter of Genesis.[10] About that chapter of Genesis a Protestant clergyman friend of mine, for many years a Navy chaplain, once sighed: if only that chapter were missing from the Bible! Although he has had that book of mine for years he still had not brought himself to reading it. Remember those who refused to look through Galileo's telescope.

Nor could the marshalling of evidence about the medieval Christian origins of science please Muslims. For them nothing is so bothersome in this age of science, that witnesses the full-scale technologization of their native lands, as are those origins. The same holds true of Hindus as well. Nehru made a laughing stock of himself in claiming that the modern scientific spirit first flourished in ancient India. As a founder of modern India, Nehru felt it had to be decorated with science too. Far more Hindu was Gandhi who decried time and again scientific preoccupations. At any rate, during one of my Oxford lectures a group of Muslims in the audience suddenly jumped up and shouted all sorts of invectives at me. One of them, raising his fist, excoriated my book, *Science and Creation: From Eternal Cycles to an Oscillating Universe.*[11] No wonder. The book deals with the most obvious but most studiedly ignored feature of scientific history: the repeated stillbirths of science in all great ancient cultures (all steeped in pantheism), and its only viable birth in the Christian West.

That viable birth was touched off by an indispensable spark, Buridan's formulation around 1348 of the idea of inertial movement. Now Buridan was part of that Christian West, where monotheism was Christian monotheism. In other words, the only God who was believed in was that God who sent his only begotten Son, in fact, created everything in Him. The Logos, this infinite rationality, therefore could create but a fully rational world, a point made by Athanasius in his struggle against the Arians. The latter, incidentally, have many followers nowadays among "enlightened" Christian theologians, busy rehabilitating Arius, the worst of all heresiarchs. But because the Son was "only begotten," that is, *monogenes, unigenitus,* He therefore dethroned from that rank the cosmos, the *to pan*, or the universe, which all educated ancient Greeks and Romans took for the only begotten emanation from the First Principle. So much about the theological matrix from which Buridan's spark jumped forth, in illustration of Christ's words: Seek first the kingdom of God and the rest will be given to you. For further details the reader of this essay may wish to search in a library for my book, *The Savior of Science*, for years now out of print.

Cultural historians of the West still need to see that spark, first noticed by Duhem, and try to digest it. But then they must part with inherited prejudices and vested interests, including plush academic emoluments, ready access to prestigious Presses, rich Foundations and the like. Why? A personal detail may not be amiss. Some months after the University of Chicago Press published my first major book, *The Relevance of Physics* (1966), I went through Chicago and paid a visit at the Press. After my visit, a big wheel there told a friend of mine on the Faculty of the University of Chicago, that "it is a pity that Father Jaki came here in a Roman collar." Why?— my friend (not a Catholic) asked in disbelief. "Because," so went the explanation, "were Father Jaki to come here in a sportjacket, we would be on our knees in front of him." I have no reason to doubt the verbatim veracity of this report.

By then the typescript of that book had been rejected by six major publishers, academic and commercial. Yet no less a figure of modern physics than Walter Heitler was to write in a review in *American Scientist* that the book is a long awaited remedy to the sundry misinterpretations of physics and should be read by all physicists. Herbert Feigl, the third reader of the typescript for the University of Chicago Press, wrote in his report that the author displays on every page an impressive scholarship. This certainly offset the puzzlement of the second reader who did not know what to do with the book.

A year after *The Relevance* was published, its first reader wrote me a long letter, to which he attached a copy of his report to the Press. He began the letter with saying that his major problem with the typescript was not to praise it too highly. But he could not resist ending his report with citing Agrippa's words to Paul: "A little more, Paul, and you will make a Christian out of me." The writer of the report disclosed to me that he, a professor of electrical engineering at a big midwestern university, was an agnostic Jew, desperately searching for the meaning of life.

In all *The Relevance* there is no pleading whatever on behalf of Jesus Christ, except perhaps a very indirect one at the end. There I quote Whitehead's phrase that only the Babe's silent birth in the manger produced in history a stir greater than science did. Why then was that professor of electrical engineering not impressed by that phrase of Whitehead, although he must have known it for years? Obviously because Whitehead, son of a Church of England clergyman, stopped believing in the Babe, just when he would have needed that faith in Him most, sometime during World War I, when a son of his became one of the war's countless victims. This loss of faith transpires everywhere in Whitehead's most widely read book, *Science and the Modern World,* which contains that priceless phrase of his.

Less than a year after *The Relevance* was published, a large envelope came to me. It contained an offprint of an article from *Life* magazine, about a new California enterprise: For a payment of five thousand dollars, (quite a sum in 1967), the enterprise would freeze my dead body, store it, and by rethawing it at a future time specified by me, would bring me back to life. The covering letter referred to my recently published *Relevance*, with the comment that it testifies to a highly intelligent mind. Perhaps, but those salesmen obviously did not read the book. Had they done so, they would not have contacted me. Half of *The Relevance* is about the basic irrelevance of physics, this most exact of all science, with respect to biology (yes, biology as dealing with the living, and not with mere molecules in motion), philosophy, theology, and basic cultural concerns. Only the other or the first half is about the restricted relevance of physics even within its own domain.

But even some very competent readers of *The Relevance* failed to grasp its message, a message striking at the heart of scientism, or the claim that the method of science is the only rational activity and whatever cannot be evaluated within its terms must be irrational, or plain bogus. Now religion, by which I emphatically do *not* mean "higher aestheticism," is bunk if scientism is valid. Hence as a believer I must have a crucial interest in showing the limited validity of the method of physics, this most exact of all sciences, precisely because it is heavily exploited and abused by scientism. To unfold that limited validity is, however, valid even if made to serve religious concerns, regardless of broader cultural considerations. In showing this limited validity or relevance, I heavily relied on statements made to that effect by physicists. Of the well over a thousand citations in

The Relevance, almost all of them are taken from physicists. Nothing is so convincing about the limited relevance of exact science than to hear it stated by physicists, old and new. The reviewer of the book in *The Atomic Scientist* had in fact to admit that "Jaki forged a powerful book." Was this an ill disguised expression of disappointment that it was not possible to say about the book that it was a forgery? Apparently it hit some in a very sensitive spot.

And why not? Had not many scientists allowed themselves to be anointed as the pontiffs of the scientific age? Not many Nobel laureates had, on receiving word about the Prize, the good sense to say something similar to the remark of my dear friend, the late Eugene Wigner. A regiment of reporters, hungry for words of wisdom about anything from the august lips of a new scientific pontiff, were lost for words on hearing him state: "The Prize did not make me a man of universal wisdom."

Hardly anything was grasped about *The Relevance* by that official of the American Scientific Affiliation when at a meeting of theirs he referred to that book as one in which "you can find all of Lord Kelvin's half-baked endorsements of the ether." Those statements take up less than two pages out of the more than six hundred. They are, of course, very educational for those who see how history, including scientific history, repeats itself. The adulation of the ether (an entity with contradictory properties) is now repeated in similar homages to the quantum mechanical vacuum, so full of energy that it allegedly performs what until recently has been ascribed to Almighty God alone: creates universes literally out of nothing. Clearly, between *that* physics and a religion distinct from *higher aestheticism* there can be no ceasefire, not a word of dialogue. Theologians, let alone their amateur brand, who think otherwise are hapless victims of confusion or ignorance or both, which is certainly a "winning" combination nowadays when coated in specious references to modern physics and the fashionable philosophies poured around it as so much toxic sauce.

To what extent this dubious technique is allowed to be practiced deserves to be illustrated by a recent example. I mean the effort of the religion reporter of the British daily, *The Independent*, to smooth over the problems created for the Church of England by its openly homosexual clergymen (and bishops). To make the farce complete the effort appeared in the formerly Catholic weekly, *The Tablet* (November 16, 1996). The writer begins with the claim of some conservatives that such and other troubles were brought upon the Church of England by its ordination of women. According to the reporter the connection is far more subtle. The subtlety he borrows from the science of quantum mechanics, of which he clearly knows only some catch-phrases: "The ordination of women acted on Anglican ideas of authority rather as an observation is supposed to act on a sub-atomic particle in quantum physics. What had once been a delicious cloud of probabilities was suddenly constrained to collapse into a measurable fact, of fixed position—if unknowable velocity." A typical runaway misuse of modern physics in support of ultramodernism in religion or something worse. Yet anyone who knows the difference between probability functions (mere ideas) and facts should realize that skullduggery is at work and not sophistication or subtlety.

But before parting with *The Relevance,* I should say something about the late Abdus Salam, who won the Nobel Prize in 1980 for his work in fundamental particle physics and who could have become a most competent reviewer of that book of mine. He wondered whether it made sense to waste excellent style over so many pages about what we all know, namely, that physics is always incomplete. Well, it seems that as a fundamental particle physicist he read only chapter 4, "The Layers of Matter," in which I stressed the chronic elusiveness of the last layer of matter. Had he read the chapter before, he would have realized that it contained a great novelty, which many Nobel laureates do not seem to know at all. More of this shortly.

Now *The Relevance* was in a sense a misnomer, though an artful one, because at that time "relevance" was *the* word that quickly caught on with everybody. As I said before, the second half of the book was about the irrelevance of physics, and only the first half about its qualified relevance even within its own domain. There I pointed out in three successive chapters that none of the three great main types of physics (organismic, or Aristotelian, mechanistic or Newtonian, and modern or essentially mathematical) can make a claim to being the final word in physics. About the two former this is now a post-mortem statement, though its historical details are enormously instructive. The death knell on the alleged finality of modern physics was sounded when it had not even completed its first quarter of a century. I mean Gödel's presentation of his incompleteness theorems before the Vienna Academy of Sciences in 1930.

Had my studies of the history physics not exposed me to the various foibles of great scientists, I would have found it impossible to understand that this theorem could fail to be applied immediately to that modern physics whose protagonists think of the world as a pattern in numbers. This application had to wait for a full generation, until it appeared in *The Relevance*. But the research that went into the writing of that book, in which only physicists, past and living, speak about physics, made it clear to me that even physicists put their trousers on one leg at a time. (Some theologians give the impression, especially when spouting scientific expressions, that they can crypto-levitate and jump into their trousers with both feet in the air).

So I was not shocked, I merely wondered, when I had my sole encounter with Professor Murray Gell-Mann. It took place at a Nobel Conference in 1976 at Gustavus Adolphus College. There he assured an audience of about 2000 strong that within a few months, but certainly within a few years, he would be able to outline the ultimate theory of fundamental particles and also prove that such a theory necessarily has to be the ultimate theory. As one of the panelists (the others were Victor Weisskopf, Steven Weinberg, Fred Hoyle, and Hilary Putnam) I could speak up before questions were taken from the floor. So I simply but firmly told Professor Gell-Mann that he would not succeed. He might find the ultimate theory, but he could be never sure that it was *the* ultimate theory, and much less that it was *necessarily* the ultimate.

Why not? he asked me in a tone which clearly revealed that he did not like to be disputed. Well, I said, in a stronger voice: because of Gödel's theorem. What theorem? he fired back, as if he had never heard of Gödel before. I suspect he had not. A few months later I gave a talk at Boston University on cosmology. I argued, among other things, that there can be no final and necessarily true cosmo-logical theory, as long as Gödel's theorems are valid. After the talk somebody from the audience walked up to me saying that I merely repeated Professor Murray Gell-Man. It was now my turn to explode: What? I asked in disbelief. Well, he said, that he had just come from Chicago where he heard Professor Gell-Man state that because of Gödel's theorems a final theory of fundamental particles cannot be constructed. I gasped and told him of my encounter with Gell-Mann a few months earlier at Gustavus Adolphus. Then it was his turn to gasp in disbelief. The punchline to this punchy story is still to come. Weinberg, still at Harvard, returned there with Hilary Putnam, who later told me that he had to give some basic information to his colleague about Gödel's theorems of which he had not been aware before he too heard me at Gustavus Adolphus. Apparently, even Hilary Putnam failed. In Weinberg's *Dreams of a Final Theory* one would look in vain for a single reference to Gödel's theorems, although they render illusory any such dream.

My life, or rather my lifelong experience concerning the relation of religion and science, has to be told largely in references to my books. In a sense they are my life-story. Ever since I started writing *The Relevance* I spent much of my working days, including Sundays, for which I beg pardon, in researching and writing, or rather writing and rewriting. That the art of writing is rewriting I learned when I just started writing *The Relevance*, or shortly after Churchill died. Then a big New York tabloid carried on its front page the facsimile of a passage from one of Churchill's famous wartime speeches. The passage, in neat typewritten form, and perfect as such, was still heavily reworked by Churchill's hand. On seeing this (later on I learned that John Henry Newman, another great master of English, usually rewrote everything three times before sending it to the printer), I got over the psychological hurdle of not being able to write a perfect copy the first try.

After that, writing has gradually become a sort of obsession, made endurable by the fact that the coming of word processors turned the otherwise tiresome business of rewriting into a relatively easy task, compared with the use of ballpoint pens and typewriters. I could not help recalling that Pierre Duhem had to write his 350 publications (including thirty vast books) with pen and ink, and with a right hand that for the last ten years of his life suffered from *crampe d'écrivain*. Often he had to hold fast his right hand with his left hand in order to continue writing. Lucky we who have lived to see the coming of PCs. They greatly increased my productivity. But, having produced so many pages for a higher purpose, namely, to strengthen those who believe in a Gospel undiluted by a "higher criticism" posing as science, I will not be threatened by that disillusion which overcame Herbert Spencer in his dying days. On seeing that his friends brought to his bedside the many books he had published, he was dismayed that he had no children of his own to stand by. He certainly might have learned a great deal from C. S. Lewis' gripping account, *A Grief Observed*.

But that productivity needed a surgical intervention to become eventually possible. In late 1953 a difficult tonsillectomy deprived me of the effective use of my voice for at least ten years. I was immediately out of all teaching and preaching. Only by writing could I go on teaching, which for me is always a preaching. I say this with no apologies or embarrassment. After having spent forty years in the academe, I find it to be the chief breeding place of a subspecies, best called spineless vertebrates. They lack intellectual spine because they refuse to admit that they preach by teaching and researching. In fact, every teaching is a sort of plain apologetics. Apologetics is pleading. To claim that one's teaching is free of even a touch of pleading on behalf of something, is to practice the art of not seeing beyond one's very nose.

I kept pleading. After publishing *The Relevance*, I published a book which was originally meant to be a chapter in it as "Physics and Psychology." It grew into *Brain, Mind and Computers*. Once this book was out of my hair (by then my hairline was rapidly receding), I could turn to what has been sheer delight to write: monographs on the history of astronomy. All had for their "ulterior" aim the illustration of chronic blindness to the obvious. First came *The Paradox of Olbers' Paradox* (1969) or the strange oversight by astronomers of the physical meaning of the darkness of the night sky. Given certain assumptions, entirely valid throughout the seventeenth, eighteenth, and nineteenth centuries, that darkness should have been seen as contradicting the assumed infinity of the universe, which became a scientific as well as scientistic or materialistic dogma in the nineteenth century.

Then came *The Milky Way* (1972), whose subtitle, *An Elusive Road for Science,* is literally true. The fact is that long before Thomas Wright hit, in 1750, upon the reason for the visual appearance of the Milky Way, it should have been spelled out by Newton and others. Worse, the explanation had to be rediscovered three times during the latter half of the 19th century. The third monograph was *Planets and Planetarians: A History of Theories of the Origin of Planetary Systems* (1976). A very checkered history indeed, especially if one considers the blind alley into which Laplace, who should have known better, guided planetary cosmogony. The material of an additional chapter, still to be written, is unfolding in the almost willful disregard of the Earth-Moon system in estimating the number of extraterrestrial civilizations. Once considerations about the improbability of our Earth-Moon system are fed into the Drake equation, the probability of a civilization with a technology similar to ours or higher to have evolved around any star in our Milky Way, would be not 10^4 but perhaps 10^{-4} or perhaps even less.

This is a point already adumbrated in my *God and the Cosmologists,* and set forth in great detail in a paper, available on the Internet, read at the October 1996 meeting of the Pontifical Academy of Sciences on micro and macro evolution. It was that meeting which John Paul II addressed with a letter on evolution, a letter that threw the media into a tailspin of their own making. The reporter of the BBC worldwide service, who contacted me in Rome, first wanted a mere five minute interview. After twenty minutes, he became disappointed on realizing that none of the headlines of the Italian newspapers were even remotely true. In sum, the pope did not embrace Darwin, while he could hail the latest achievements in evolutionary science as distinct from Darwinian ideology, the latter having for its basic dogmas the uncreatedness of the universe, and the perishing of the soul with bodily death. Whatever the sins of the Italian press, it is hardly a virtue that the BBC did not find the very truth about what the pope said worth broadcasting.

Here belong a few words about my translation into English of three classics of the history of astronomy. First, Giordano Bruno's *Ash Wednesday Supper* (1975), for which Bruno would have deserved to be burnt by scientists, that is, all the fledgling Copernicans, right there and then when he published it in 1584. Then came Lambert's *Cosmological Letters* (1976), which prompted its reviewer in *Scientia* (Milan) to say that its notes, almost as many pages as the book itself, would necessitate the rewriting of the history of cosmology during the 17th and the first half the 18th centuries. The third was Kant's *Universal Natural History and Theory of the Heavens* (1981). Doing this gave me a particular delight. The book is only one third Kant. Equally long is my introduction and the notes added to it. They show that in speaking of science Kant was an artful poseur. The evidence may perhaps awaken those who keep Kant, the philosopher, in high esteem because of his alleged expertise in science. Cutting science and scientists, but above all some "scientific" philosophers to size, would help create a clear atmosphere in which talk about religion might not go off, again and again, on a tangent and end in clever games with mere ideas.

Sometime before all this, actually in 1970 and 1971, I wrote *Science and Creation,* mentioned above. While the secularists groaned (plenty of them among historians of science), some religionists —among them the reviewer in the *Bulletin* of the Victoria Institute, London— were ecstatic. Certainly if one is clear-sighted enough to know that pantheism is the only logical alternative to biblical revelation culminating in Christ, nothing is so unwelcome in this scientific age as the word that pantheism was the cause of the stillbirths of science and that only a Christ-based monotheism made science experience its only live birth.

The next item I shall talk about is my Gifford Lectures, given at the University of Edinburgh in 1974-75 and 1975-76, published as *The Road of Science and the Ways to God*. (That I was invited has much to do with the interest which my friend, Thomas F. Torrance, took in my writings, an interest touched off by his reading of *The Relevance*). Giving a long title to my Gifford Lectures certainly taught me that it does not easily stick, however true to the content of the book. I should have perhaps called it "Science and the Ultimate" and it still would have remained true to its real thrust. I argued in those twenty lectures that a scientist's or a philosopher's choice of the ultimate in intelligibility and being determines his dicta on science, for better or for worse. Further, one can choose God as the ultimate, only if one's epistemology permits the cosmological argument, an argument very different from Kant's distortions of it, couched partly in his inept references to science. It was only fifteen years later that I had the opportunity to argue in detail, in the course of my Forwood Lectures at the University of Liverpool, that while science is unable to prove the reality of the universe, philosophy, and philosophy alone, can do it. Such is the gist of that series of lectures, *Is there a Universe?* Therefore scientific cosmologists should rename their subject matter as supergalactology or something. Such was at least the inference drawn by a prominent British astronomer who read the book. Most other astronomers still have to discover that they gently cheat in writing books on cosmology.

The Templeton Prize of 1987 had something to do with that invitation to Liverpool. Before that the Stillman Chair at Harvard could have been mine but for the asking. I declined. I stuck with my University, Seton Hall, simply because I do not think that Catholic priests should abandon Catholic Universities for places more prestigious in the eyes of the world. A Catholic priest should not lose sight of the biblical warning that a man of God must be on guard against abandoning the lean in favor of the fat. Such a hint is hardly a hint to the liking of liberal Catholic intellectuals.

Whatever my failures to live up to the ideals of the priesthood, I have perhaps explicitly served it through this or that book of mine. The Gifford Lectures prompted the conversion to Catholicism of a prominent midwestern industrialist. Many others wrote about the spiritual profit they derived from it. The Church was directly the object of three books of mine. The first, *Les tendances nouvelles de l'ecclésiologie*, grew out of my doctoral dissertation. It was reprinted during Vatican II. The *And on this Rock,* now out in its third enlarged edition, is about a rock in Cacsarea Philippi, about the word rock in the Old and New Testament, and about the unfailing preaching by the papacy of Jesus' divinity. *The Keys of the Kingdom* takes its starting point from the history of key-making. Science, if properly used, can be of great assistance in building a theological argument, even in biblical theology.

This is further illustrated by two other books of mine. One, *Genesis 1 through the Ages*, has already been mentioned, though not the fact that the first chapter of the Bible is also its most misinterpreted chapter. I would not have written it, had I not surmised that the explanation which does justice both to Bible and science must be sought in reference to the sabbath observance. This point is now explicitly treated at some length in my essay, "The Sabbath Rest of the Maker of All," in the *Asbury Journal of Theology*.[12] The other book to be mentioned is *Bible and Science*.[13] The science in question is not archeology, but hard physical science. I argue there, among other things, that one's reading of the Bible should first be disentangled from the false opposition between the Hebrew and Greek mind, if one is to grasp the biblical authors' commonsense perception of reality. This plainly epistemological point (any attempted refutation of which implies implicit trust in such a perception) bears heavily on the trustworthiness of the biblical authors' reports on miracles. The same holds true of any plain witness about miraculous events, such as the cures at Lourdes. As I argued in my little book, *Miracles and Physics* (1990), nothing is more mistaken than to invoke Heisenberg's uncertainty principle as if it provided a chance for God to do something physical without interfering with the laws of physics. The theologian, who is still interested in free will, will make an equally great mistake when he tries to defend it with a reliance on quantum mechanics. A long essay of mine, "Determinism and Reality" (1990) will make all this clear.[14]

Following the Gifford Lectures, I had the honor to give in 1977 the Fremantle Lectures at Balliol College in Oxford. They appeared as *The Origin of Science and the Science of its Origin.* It is an analysis of theories on the origin of science from Bacon on to the present. It should be read as a companion volume to *Science and Creation* and the Gifford Lectures. It was at about that time, or shortly before it, that I was elected a permanent Visiting Fellow at Corpus Christi College. By then I had become a good friend of Dr. Peter Hodgson, at Corpus, a friendship that time and distance has not weakened.

It was he who called the attention of the Farmington Institute of Oxford to my work. This resulted in two series of lectures, given in 1988 and 1989, respectively. The former came out as *God and the Cosmologists* (1989), the latter as *The Purpose of it All* (1990), both in collaboration with Scottish Academic Press, whose director, Dr. Douglas Grant, has shown an unflagging interest in my work ever since he published *Science and Creation.*

In my history of fighting the good fight on behalf of good science and true religion, I should not forget my encounter with Mr. Chauncey Stillman and the Wethersfield Institute he founded. This Institute sponsored the series of lectures that came out as *The Savior of Science* (1988). There I gave a full-bodied treatment to the theses, already mentioned above, about the crucial role which a monotheism steeped in belief in the divinity of Christ played in the fate and fortune of science, and that only such a belief shall provide moral strength for a proper use of the tools of science that have an increasingly frightening range.

In *The Savior of Science* I also dealt with evolution. In doing so I recalled that passage about water spiders, quoted above to illustrate a crucial point in coping with Darwinism, that strange mixture of incomplete science and complete philosophical obfuscation. The passage about water spiders must not be used as a proof that God created them specially and millions of other species, all of which display "skills" that Darwinism never explained scientifically. By that scientific explanation I mean the step by step demonstration that countless modifications of an organ did indeed lead to an organ very different and used for very different purposes. The great strength of Darwin lies in his bringing together powerful indications that allow only one inference: all species have arisen in strict dependence on the alteration of other species. Christians, who believe in an Almighty

God, must be grateful to Darwin for having, however unwittingly, reminded them, that it is not the materialists who must have the strongest confidence in the unlimited power of matter about everything material, but the ones who believe in the Almighty Creator of all matter.

Darwinists can only be pitied for their often unabashed materialism. On its basis one cannot even infer that generalizations, which are their daily food, such as species, genera, phyla, kingdom etc, are really valid ones. Even more to be pitied are those (and most Darwinists are such) who devote their entire life to the purpose of proving that there is no purpose. To prove purpose, which in man goes together with free will, let alone the analogical realization of purposeful action in animals and plants, one needs an epistemology, which, among other things, is incompatible with easy recourses to the Bible, but, as I argued in *Bible and Science*, still shines through every page of it.

So much in a way of background for the remark which in *The Savior of Science* follows another instance of spiders if not water spiders:

> There is a Queensland spider called *the magnificent*, because of the fine colouring of the female. But it is her way of catching moths that concerns us at present. She hangs down from a line and spins a thread about an inch and a half in length, bearing at the free end a globule of very viscid matter, a little larger than the head of a pin. The thread is held out by one of the front legs, and on the approach of a moth the spider whirls the thread and the globule with surprising speed. The moth is attracted, caught, pulled up, killed, and sucked. When it is touched by the whirling globule it is helpless as a fly on fly-paper. We may well say of the magnificent spider, *c'est magnifique*.

Here I remarked that the continuation of this *"c'est magnifique"* should not be *mais ce n'est pas la guerre*, whatever the endless skirmishes provoked by the illogicalities of Darwinism, but, rather, *ce n'est pas la sélection naturelle*. This is not to suggest that natural selection does not play a very important role. But in saying this one heavily relies precisely on the kind of philosophy which most champions of evolution claim to have eliminated by their science. They carefully sweep under the rug the fact that nobody has yet demonstrated *scientifically* that natural selection does indeed produce the magnificence of the behavior of water spiders and countless other creatures. Those who claim that such a demonstration is on hand are as mistaken as those who despise evolutionary biology because it has to rely heavily on generalizations and extrapolations. But so does religion. Whenever a Christian apologist grows unmindful of this, he pulls the very ground from under his very feet. There are theological counterparts of those who no less than the purposive debunkers of purpose illustrate the darkening of the intellect, one of the chief effects of man's erstwhile Fall.

Since I began with Duhem, I perhaps should close with remarks about other works of mine connected with him. Providence (misnamed nowadays as Chance, writ large) brought me into contact around 1985 with Norbert Dufourcq of Paris. He was a prominent musicologist, the son of the famed church historian, Albert Dufourcq, who in turn was one of Duhem's best friends at the University of Bordeaux. Norbert Dufourcq thus inherited much of Duhem's correspondence as well as the correspondence of Hélène, Duhem's daughter. This connection was the source of the material which appeared in a book, *Reluctant Heroine: The Life and Work of Helene Duhem* (1991).[15] Her story, a very gripping one, should be of interest not only to the professional historian of science, but also to anyone wishing to know the extent to which prominent publishing firms are willing to ignore their contractual obligations and serve thereby powerful ideological interest groups in the academia who just simply cannot contemplate the airing of some facts.

It was also through Providence that I met Mlle Marie-Madeleine Gallet, Albert Dufourcq's niece, who was Hélène Duhem's great support in her closing years. It was from Mlle Gallet that I obtained for publication Duhem's albums of landscape sketches. A vast selection of them appeared, with my introduction, as *The Physicist as Artist: The Landscapes of Pierre Duhem* (1990).[16] The Académie des Sciences in Paris let me use its collection of the almost daily letters which Pierre Duhem wrote to his daughter between 1909 and 1916. A selection from them appeared, with my notes and introduction, as *Lettres de Pierre Duhem à sa fille Hélène*. All this material greatly helped me in writing *Scientist and Catholic: Pierre Duhem* that came out in French and in Spanish as well.

The full list of my publications until 1990 is available in *Creation and Scientific Creativity: A Study in the Thought of S. L. Jaki* by Paul Haffner, originally a doctoral dissertation that earned its author *summa cum laude* at the Gregorian University at Rome.[17]

There one can read my statement upon my being inducted, in 1991, into the Pontifical Academy of Sciences as one of its honorary members. There I paid my homage to Duhem and also voiced my view that what God had put into separate conceptual compartments, no one should try to fuse together.

I hope and pray that God will give me the strength to write a summary of my views on science and religion, the thrust of which may easily be gathered from this brief essay. Another future plan of mine is now under way, under the title, *A Treatise on Truth,* in which I plan to summerize my philosophy, that is, epistomology and metaphysics. There I will make much of what I have learned from Etienne Gilson about methodical realism, from the Liverpool philosopher, J. E. Turner, about the logical fallacy of claims that the uncertainty relation disproved causality. There I will develop further the unique status of the category of quantities among all categories, by which I understand the ten Aristotelian categories and not their Kantian counterfeits. A foretaste of this can be found in my latest publication, a long essay, entitled, "The Limits of a Limitless Science," that first appeared in Italian translation in *Con-Tratto,* as commissioned for its 1996 Yearbook, but will soon appear in English as well in the *Asbury Theological Journal.*

While all talk about religion (theology) stands or falls with the philosophy it rests upon, this is less true of science. The more exact is a science, such as physics, the more its conclusions become independent of the philosophical matrix out of which they have grown. For insofar as those conclusions are quantitative, they have a validity independent of the philosophy which the individual scientist tags on them. Thus science can be compared to the building of an edifice: The completed theory is like an edifice from which all the scaffolding (including philosophy) has been removed. The edifice, however, contains nothing philosophical. It is a mere structure in numbers. It is in this sense that one should take Hertz's famous dictum: "Maxwell's theory [of electromagnetism] is Maxwell's system of equations," a dictum that I cannot repeat often enough. Nothing remotely as fundamental has ever been said by a great physicist about the physical theory of an even greater physicist.

When cut to bare bones, exact science is nothing more, nothing less than a system of equations. There would be no conflict whatever between science and theology were scientists truly mindful of this truth. But scientists are, like all of us, philosophers as well. The only way to avoid philosophy is to say nothing. The trouble is that nothing can sell a bad philosophy more effectively than attaching it to a splendid science. (Thus science is turned into one of the three S's of modern life: Sports, Sex, Science, all writ large.) The converse is not true; no amount of science, insofar as it is science and not something more, can justify a single philosophical proposition and much less a single theological statement, which has to be a proposition not about how the heavens go, but how to go to Heaven. Unfortunately, theologians, believing themselves to be in possession of eternal truths, are prone to discourse about mere temporalities, such as the physical universe, about whose measures, large and small, science is the sole arbiter.

In sum, ever since science obtained its only viable birth, it keeps unfolding the enormous potentialities of its basic equations of motion. Thus science reveals more and more about the quantitative aspects of all things in relative motion to one another. In performing this revelatory function, science makes empirically, that is, measurably verifiable predictions. To perform this function science, once truly born, needs no extraneous revelation. This is why science is neither theistic, nor atheistic; it is just science, unlike theology which has to be theistic. Theology, unless it wants to degenerate into a branch of mere religious studies, must be, at the very minimum, about a personal God, who can and must be worshiped, and not merely admired as a superior form of sunrise or sunset, or a mere mist hovering over a well-manicured lawn.

Sundays are not for communing with Nature, writ large, but for the worship of a personal God, who has absolute sovereignty even over that mankind that makes the greater mess of itself the more it lives up to its declaration of total autonomy. Modern man, so proud of his science, still needs to learn that science reveals precisely the fact that the universe is restricted to do, relatively speaking, very few things. A little reflection on this may prompt any clear-thinking to draw the conclusion and genuflect. For the answer to the question why is the universe such, and not something else, leads one to the ultimate cause of any suchness, or the source of all, which is Almighty God. He could have created an infinite variety of worlds. The one in existence is the result of His sovereign creative choice. This is why the actually existing universe has a stunning set of specificities whose investigation is our great intellectual stepping

stone to the recognition of the One who is existence himself. For as
He revealed himself, He is the One who IS.

[1] H. Hertz, *Electric Waves*, tr. D. E. Jones (1893; New York: Dover, 1962),
p. 21.

[2] A. S. Eddington, *Science and the Unseen World* (New York: The
Macmillan Company, 1930), p. 58.

[3] J. Arthur Thomson, *Biology for Everyman* (New York: E. P. Dutton, 1935),
p. 360,

[4] S. L. Jaki, *Is There a Universe?* (Liverpool: Liverpool University Press,
1993).

[5] S. L. Jaki, *God and the Cosmologists* (1989; 2d rev. ed; Edinburgh:
Scottish Academic Press; Royal Oak, MI: Real View Books, 1998).

[6] S. L. Jaki, *The Road of Science and the Ways to God* (Chicago: University
of Chicago Press; Edinburgh: Scottish Academic Press, 1978).

[7] S. L. Jaki, "Words: Blocks, Amoebas, or Patches of Fog? Artificial
Intelligence and the Foundations of Fuzzy Logic," *Proceedings of the
International Society for Optical Engineering*, vol. 2761 (1996), pp. 138-43.

[8] P. Duhem, *The Aim and Structure of Physical Theory* (Princeton:
Princeton University Press, 1954), pp. 273-74.

[9] S. L. Jaki, *Uneasy Genius: The Life and Work of Pierre Duhem*
(Dordrecht: Nijhoff, 1984).

[10] 2d enlarged ed.; Royal Oak, MI: Real View Books, 1998).

[11] S. L. Jaki, *Science and Creation: From Eternal Cycles to an Oscillating
Universe* (1974; 2d enlarged ed.; Edinburgh: Scottish Academic Press, 1987).

[12] 50 (Spring 1995), pp. 37-49.

[13] S. L. Jaki, *Bible and Science* (Front Royal, VA: Christendom Press,
1996),

[14] S. L. Jaki, "Determinism and Reality," in *Great Ideas Today 1990*
(Chicago: Encyclopedia Britannica, 1990), pp. 277-301.

[15] S. L. Jaki, *Reluctant Heroine: The Life and Work of Hélène Duhem*
(Edinburgh: Scottish Academic Press, 1991).

[16] S. L. Jaki, *The Physicist as Artist: The Landscapes of Pierre Duhem*
(Edinburgh: Scottish Academic Press, 1990).

[17] P. Haffner, *Creation and Scientific Creativity. A Study in the Thought of
S. L. Jaki* (Front Royal, VA: Christendom Press, 1991).

Index of Names

Agassiz, L., 96-97
Aiton, E. J., 178
Alexander, S., 3, 144
Ambartsumian, V. A., 63, 103, 138
Anaxagoras, 35
Aquinas, *see* Thomas, Aquinas
Arendt, H., 71-72
Aristarchus, 28
Aristotle, 14-17, 36, 51, 85, 97-98,
  110-11, 114, 185, 221
Arius, 165, 225
Arnold, M., 193-94, 201
Ashari, al-, 167
Athanasius, Saint, 218
Augustine of Hippo, Saint, 2, 52, 123
  Averroes, 167
Avicenna, 167
Ayer, A. J., 18, 90

Babbage, C., 40
Bacon, F., 35
Bahcall, J. N., 62, 75
Baxter, W., 164
Becker, C. L., 69-70, 76
Bell, J. S., 169
Bergson, H., 71, 112, 177, 200
Berkeley, G., 36
Bernard, C., 10
Berra, D., 88
Berra, Y., 88
Blackmore, J. T., 179
Bohr, N., 38, 61, 95, 134, 147, 168-
  69, 172, 178, 186-87
Bondi, H., 147
Born, M., 133, 168, 172, 185, 199

Brahe, T., 28, 58-59
Brawer, R., 147
Browne, M. W., 146, 179
Bruno, Giordano, 29, 59, 233
Brush, S. C. 179
Buckley, W. F., 163
Buridan, J., 28, 51, 225
Burke, E., 177
Bury, J., 73
Butler, N. M., 207

Calame, O., 29
Calvin, J., 224
Cantor, G., 115
Capra, F., 178
Carnap, R., 22, 99, 122
Carroll, L., 88, 132
Chalmers, D. J., 23
Chandrasekhar, S., 195
Chesterton, G. K., 41, 174
Churchill, W., 194, 231
Comte, A., 37, 60, 75, 110, 171, 173,
  179, 182
Condorcet, M. J., 200
Cooper, D., 194, 201
Copernicus, N., 28, 39, 52, 72, 74, 92
Copleston, F., 116
Crossan, J. D., 177
Curtius, E., 201
Cusa, Nicholas of, 171

Darwin, C., 14, 52, 55, 67, 95-96,
  124, 210, 233, 236-37
Darwin, G. H., 26
Davies, P. C. W., 134

(continued from p. ii)

By the same author

The Only Chaos and Other Essays

The Purpose of It All
(Farmington Institute Lectures, Oxford, 1989)

Catholic Essays

Cosmos in Transition: Studies in the History of Cosmology

Olbers Studies

Scientist and Catholic: Pierre Duhem

Reluctant Heroine: The Life and Work of Hélène Duhem

Universe and Creed

Genesis 1 through the Ages

Is There a Universe?

Patterns or Principles and Other Essays

Bible and Science

Theology of Priestly Celibacy

Means to Message: A Treatise on Truth

God and the Sun at Fatima

Newman's Challenge

* * *

Translations with introduction and notes:

The Ash Wednesday Supper (Giordano Bruno)

*Cosmological Letters on the Arrangement
of the World Edifice* (J.H. Lambert)

Universal Natural History and Theory of the Heavens (I. Kant)

Note on the Author

Stanley L. Jaki, a Hungarian-born Catholic priest of the Benedictine Order, is Distinguished University Professor at Seton Hall University, South Orange, New Jersey With doctorates in theology and physics, he has for the past forty years specialized in the history and philosophy of science. The author of almost forty books and over a hundred articles, he served as Gifford Lecturer at the University of Edinburgh and as Fremantle Lecturer at Balliol College, Oxford. He has lectured at major universities in the United States, Europe, and Australia. He is honorary member of the Pontifical Academy of Sciences, *membre correspondant* of the Académie Nationale des Sciences, Belles-Lettres et Arts of Bordeaux, and the recipient of the Lecomte du Nouy Prize for 1970 and of the Templeton Prize for 1987.